Human–Computer Etiquette

Cultural Expectations and the Design Implications They Place on Computers and Technology

SUPPLY CHAIN INTEGRATION
Modeling, Optimization, and Applications

Sameer Kumar, Series Advisor
University of St. Thomas, Minneapolis, MN

Human-Computer Etiquette: Cultural Expectations and the Design Implications They Place on Computers and Technology
Caroline C. Hayes and Christopher A. Miller
ISBN: 978-1-4200-6945-7

Closed-Loop Supply Chains: New Developments to Improve the Sustainability of Business Practices
Mark E. Ferguson and Gilvan C. Souza
ISBN: 978-1-4200-9525-8

Connective Technologies in the Supply Chain
Sameer Kumar
ISBN: 978-1-4200-4349-5

Financial Models and Tools for Managing Lean Manufacturing
Sameer Kumar and David Meade
ISBN: 978-0-8493-9185-9

Supply Chain Cost Control Using Activity-Based Management
Sameer Kumar and Matthew Zander
ISBN: 978-0-8493-8215-4

Human–Computer Etiquette

Cultural Expectations and the Design Implications They Place on Computers and Technology

Edited by

Caroline C. Hayes • Christopher A. Miller

CRC Press is an imprint of the
Taylor & Francis Group, an **informa** business
AN AUERBACH BOOK

Auerbach Publications
Taylor & Francis Group
6000 Broken Sound Parkway NW, Suite 300
Boca Raton, FL 33487-2742

Printed in the United States of America on acid-free paper
10 9 8 7 6 5 4 3 2 1

International Standard Book Number: 978-1-4200-6945-7 (Hardback)

Library of Congress Cataloging-in-Publication Data

Human-computer etiquette : cultural expectations and the design implications they place on computers and technology / edited by Caroline C. Hayes, Christopher A. Miller.
p. cm.
Includes bibliographical references and index.
ISBN 978-1-4200-6945-7 (hardcover : alk. paper) 1. Computers--Social aspects. 2. Human-computer interaction. 3. User interfaces (Computer systems) 4. Etiquette. I. Hayes, Caroline. II. Miller, Christopher Allan.

QA76.9.C66H823 2010
005.4'37--dc22 2010044376

Visit the Taylor & Francis Web site at
http://www.taylorandfrancis.com

and the Auerbach Web site at
http://www.auerbach-publications.com

This book is dedicated to our mentors, students,
co-workers, friends, and especially our family
members: Marlene, Michael, and Andrea.

Contents

Photo and Figure Credits

Cover Photo: Courtesy of Nancy Gail Johnson
Section 1: Courtesy of Nancy Gail Johnson
Section 2: Courtesy of Nancy Gail Johnson
Section 3: Courtesy of RIKEN-TRI Collaboration Center for Human-Interactive Robot Research, Nagoya, Japan
Section 4: Courtesy of Alan Rogerson
Section 5: Courtesy of Caroline C. Hayes
Section 6: Courtesy of Jared von Hindman, Headinjurytheater.com

Acknowledgments

This book could not have come together without the help of many people. We would like to thank our reviewers who volunteered their time to comment on the chapters as they developed. We would also like to thank the editorial staff at the Taylor & Francis Group, without whose support and encouragement this book could not have come together. These people include Sameer Kumar, who first encouraged us to develop our ideas into a book, Maura May, and Raymond O'Connell, who supported and encouraged us during the process. Sadly, Raymond O'Connell did not live to see the finalization of the book. He dedicated most of his life to publishing, and we proudly include this book as one of his final accomplishments. We would like to give a special thanks to Victoria Piorek for all of her administrative assistance at all phases of development of the book, which made our work possible.

The Editors

Caroline Clarke Hayes is a professor at the University of Minnesota with appointments in mechanical engineering, computer science, and industrial engineering. Her research focuses on finding ways to make people and technology work together more effectively and harmoniously. Current projects include decision support tools that can augment peoples' own decision-making approaches, as well as tools for improving team collaboration over distance. She currently heads the graduate program in human factors, and is serving as the chair of the Provost's Women's Faculty Cabinet at the University of Minnesota.

Christopher A. Miller is the chief scientist and co-owner of Smart Information Flow Technologies (SIFT), a small Minneapolis-based research and development business specializing in human factors, computer science, and human–computer interaction. His interests include human–automation interaction, computational models of social interaction, and human performance and interpersonal and intercultural communication. His current work includes multicultural politeness models for training and cultural interpretation, human information flow models, and work on human interaction with multiple uninhabited vehicles. Prior to his involvement with SIFT, he was a research fellow with Honeywell Laboratories.

Contributors

Hall P. Beck
Appalachian State University
Boone, North Carolina

Timothy Bickmore
College of Computer and
Information Science
Northeastern University
Boston, Massachusetts

James P. Bliss
Old Dominion University
Norfolk, Virginia

Annabelle Boutet
Telécom Bretagne
Brest-Iroise, France

Gilles Coppin
Telécom Bretagne
Brest-Rennes, France

Tulio Rojas Curieux
Universidad del Cauca
Calle, Popayán, Colombia

Michael C. Dorneich
Honeywell Laboratories
Minneapolis, Minnesota

Mark Dzinsolet
Cameron University
Lawton, Oklahoma

Harry Funk
Smart Information Flow
Technologies (SIFT)
Minneapolis, Minnesota

Rego Grandlund
Rationella Datatjänster
Rimforsa, Sweden

P.A. Hancock
Department of Psychology
Institute for Simulation and Training
University of Central Florida
Orlando, Florida

Caroline C. Hayes
Department of Mechanical Engineering
University of Minnesota
Minneapolis, Minnesota

W. Lewis Johnson
Alelo Inc.
Los Angeles, California

David B. Kaber
Edward P. Fitts Department of Industrial & Systems Engineering
North Carolina State Univesrity
Raleigh, North Carolina

Helen Altman Klein
Wright State University
Dayton, Ohio

Jason Kring
Embry-Riddle Aeronautical University
Daytona Beach, Florida

Mei-Hua Lin
Jalan Universitiy
Petaling Jaya, Selangor, Malaysia

Ida Lindgren
Linköping University
Linköping, Sweden

Katherine Lippa
Austin, Texas

Tong Liu
Institute of Information and Mathematical Sciences (IIMS)
Massey University
Auckland, New Zealand

Donald R. Lampton
Army Research Institute, Orlando Field Unit
Orlando, Florida

Santosh Mathan
Honeywell Laboratories
Redmond, Washington

Christopher A. Miller
Smart Information Flow Technologies (SIFT)
Minneapolis, Minnesota

Raja Parasuraman
George Mason University
Fairfax, Virginia

Linda G. Pierce
Army Research Institute
Aberdeen, Maryland

Franck Poirier
Universite de Bretagne-Sud
Vannes, France

Santiago Ruano Rincón
Telécom Bretagne
Brest, France

Kip Smith
Cognitive Engineering and Decision Making, Inc.
Des Moines, Iowa

Patricia May Ververs
Honeywell Laboratories
Columbia, Maryland

Vanessa Vikili
Smart Information Flow Technologies (SIFT)
Minneapolis, Minnesota

Ewart de Visser
George Mason University
Fairfax, Virginia

Ning Wang
Institute for Creative Technologies
University of South California
Los Angeles, California

Stephen Whitlow
Honeywell Laboratories
Minneapolis, Minnesota

Brian Whitworth
Institute of Information and Mathematical Sciences (IIMS)
Massey University
Auckland, New Zealand

Peggy Wu
Smart Information Flow Technologies (SIFT)
Minneapolis, Minnesota

Tao Zhang
Edward P. Fitts Department of Industrial & Systems Engineering
Vanderbilt University
Nashville, Tennessee

Biwen Zhu
Edward P. Fitts Department of Industrial and Systems Engineering
North Carolina State University
Raleigh, North Carolina

1

Human–Computer Etiquette

Should Computers Be Polite?

CAROLINE C. HAYES AND
CHRISTOPHER A. MILLER

Contents

> **1st Rule** Every action done in company ought to be done with some sign of respect to those that are present.
>
> **—George Washington's Rules of Civility***

Why write a book about human–computer etiquette? Is etiquette a concept that is relevant when dealing with things that are not human—that are not even living beings? Typically, we think of etiquette as the oil that helps relationships run smoothly and soothes human feelings. The use or absence of etiquette impacts how participants feel about the interaction, their likelihood of complying with requests, and the quality of the long-term relationship between participants. However, computers do not have feelings and do not "feel" their impact on others. Computers have not traditionally concerned themselves with relationships, therefore why should etiquette be meaningful in interactions with computers? Clearly it is because the *users* of computers have feelings; and etiquette, even coming from a computer, impacts how users trust, accept, and interact with even a mechanical device.

* Washington, George (2003) George Washington's Rules of Civility, fourth revised Collector's Classic edition, Goose Creek Productions, Virginia Beach, VA.

One might say that etiquette forms a critical portion of the "rules of engagement" for social interactions between autonomous parties in any rich, and therefore ambiguous, setting.

We argue that etiquette is not only *relevant* for understanding interactions between people and computers, it is *essential* if we are to design computer assistants that can work effectively and productively with people (Miller, 2000; 2004). Regardless of whether computers are designed to exhibit etiquette, human users may interpret their behavior as polite or rude. However, we first need to understand more about what etiquette is in human interactions, and what is its function. How do we view computers? And what is their role in our society?

1.1 Eitquette: Definition and Role

Etiquette is typically thought of as a set of socially understood conventions that facilitate smooth and effective interactions between *people*. Interactions may involve spoken, written, or nonverbal communications. They include far more than words; meaning comes from a combination of many channels such as tone or voice, facial expressions, body language, and actions such as holding a door or offering the correct tool before it is requested. Brown and Levinson (1987) state that "politeness, like formal diplomatic protocol ... makes possible communication between potentially aggressive parties." Politeness is essential in establishing and building relationships and trust, turning potentially aggressive parties into cooperative parties. It helps people to live and work effectively together, and to coordinate their actions as members of a productive group in their daily lives. In summary, etiquette underlies "the foundations of human social life and interactions."

Etiquette is not always about being pleasant, it is about being *appropriate*—behaving in a way that others will understand and perceive to be correct in context. Insufficient politeness or inappropriate interruptions are viewed as rude. However, being overly polite may be viewed as obsequious and therefore irritating, and failing to interrupt in an emergency ('Fire!") can be downright dangerous. Etiquette can be used to help one to be seen as "nice" or "polite," but it may also be used to communicate emotions that are less pleasant such as dissatisfaction, uncertainty, urgency, and prohibition.

Likewise, we define *human–computer etiquette* as a similar set of conventions that facilitate smooth and effective interactions between people and computers. For computers to be successful in etiquette, they must produce nuanced responses that are sensitive to the context and reactions of people. The reasons that computers need to be appropriately polite are very pragmatic; like people, computer agents need to exhibit appropriate etiquette if they are to be accepted as part of a working team, gain trust from their human collaborators, and enhance rather than disrupt work and productivity. For example, if a computer assistant interrupts an airline pilot during landing, the consequences could be disastrous. Even in a less life-critical task such as word processing, if a computer assistant interrupts at an inappropriate time, the person interrupted will likely feel irritation, their concentration may be disrupted, and their productivity reduced. If computer assistants are to be viewed as valuable team members rather than "bad dance partners," they must be designed to follow the rules of good team players.

Two of the major etiquette challenges are that (1) many of the conventions of etiquette are implicit, and (2) etiquette is highly dependant on culture and context. These are equally true for people attempting to behave appropriately in social situations, and for software designers who endeavor to build software that behaves appropriately. There are some etiquette conventions that may be stated explicitly, for example in *George Washington's Rules of Civility*, Emily Post's *Etiquette*, or *Miss Manners' Guide to Excruciatingly Correct Behavior* (Washington, 4th ed., 2003; Post, 17th ed., 2004; Martin, 2004).* However, etiquette is more often implicitly and unconsciously understood and applied. People learn etiquette conventions over a lifetime, often without consciously realizing they are learning specific social interactions. Etiquette goes far beyond these explicit sets of rules and protocols. This can make it challenging for people attempting to learn the conventions of other cultures, or for software designers attempting to build those conventions into explicit computer systems.

* Each of these are American etiquette guides from the 18th, 20th, and 21st centuries. The first is derived from an earlier 16th century French etiquette guide, *Bien-séance de la Conversation entre les Hommes* (*Good Manners in Conversation with Men*).

1.2 Etiquette Is Situated in Culture, Time, Place, and Context

Etiquette is both a reflection and expression of culture. Culture changes with geographic location, and over time; it is not static. The following "rules" from "George Washington's Rules of Civility," written by Washington when he was a boy in the 1700s, illustrate both how American culture has changed over time, and how much of our current etiquette is implicit.

26th Rule: "In pulling off your hat to persons of distinction, such as noblemen, justices, churchmen & etc., make a reverence; bowing more or less according to the custom of the better bred and quality of persons. ..."

27th Rule: "Tis ill manners to bid one more eminent than yourself to be covered [e.g., to put your hat back on] as well as not to do it to whom it's due. Likewise, he that makes too much haste to put on his hat does not well, yet he ought to put it on at the first or second time of being asked."

To a current day American, these rules appear not only irrelevant, but exhausting. How could one possibly keep track of when to bow, who should take off their hat for whom, and precisely how long one should keep it off? These rules appear irrelevant because American culture no longer places as strong an emphasis on social status or acknowledging its associated rituals.

100th Rule: "Cleanse not your teeth with the tablecloth, napkin, fork or knife ..."

The 100th rule appears laughable now because no one would ever think to write it down as it is implicitly understood in American culture that this is simply not done. One would probably not find the equivalent of this rule in etiquette guides such as those by Emily Post or Miss Manners. However, it is exactly this type of culturally implicit convention that makes it difficult for people to know how to behave in other cultures, or for programmers to know how to instruct a computer in what to do. Issues of culture and etiquette will be explored in more depth in Chapters 2 through 6.

1.3 Computers: Machines or Hybrid Beings?

Not only do people expect computers to follow many of the rules of social interaction, but the reverse is also true; they frequently treat computers as if they were social beings, despite being fully aware that they are not (Figure 1.1). For example, people poured their hearts out to ELIZA, the computer program that mimicked a Rogerian psychiatrist, even when they knew it was a machine (Weizenbaum, 1966). Nass found that people respond to computers in much the same ways as they respond to people, along many dimensions (Nass et al., 1994). For example, people tend to discount it when people praise themselves, "I am the world's expert on rainforests," but take it more seriously when other people praise them, "She really is the world's expert on rainforests." Nass performed an experiment in which he asked subjects to listen to computer voices "praising" a computer tutor, then fill out evaluations of the tutor's competence. Their evaluations showed that people tended to discount statements of praise that came from the same workstation that was also running the tutor, but they took the praise more seriously if it came from a different computer. Thus, it appeared that they treated praise from a computer in much the same way as they would treat praise from a person.

Nass also found that people have social responses to computers regardless of whether they have human-like features or not (Nass et al., 1994). However, adding human-like features may enhance their tendency to anthropomorphize. For example when computers are

Figure 1.1 People often respond to computers in much the same way as they would to other people. (Courtesy of Kolja Kuehnlenz © 2009 Inst. of Automatic Control Engineering (LSR), TU Munich).

given human voices, faces, or forms, people tend to apply the same gender and race stereotypes to the computer as they would to a person of that gender or race (Nass et al., 1997; Moreno et al., 2002; Gong, 2008). Thus, it appears that people have socially-based expectations of how computers should behave, whether they are conscious of these expectations or not. These observations reflect a long standing, ambiguous relationship between people and computers, and more generally between people and technology. People know that computers are "things" yet they treat them as if they were something more.

The human tendency to treat machines as more than machines is not isolated to computers (in all their various forms, from laptops to robots); people also form attachments and relationships with other types of artifacts, and respond to them in ways that are essentially social. (Indeed, cf. Dennett, 1987) For example, many people talk to their cars, give them names, pat their dashboards encouragingly, and curse at them when they fail. They may form human-like attachments to tools that are an essential part of their work and creative processes. The concert cellist Jacqueline du Pre was depicted in a memoir as treating her cello unkindly when she felt the demands of her musical career were limiting her life; she left the valuable antique cello on the hotel balcony in the snow, and "forgot" it several times in the back of cabs. Then when her life started to look up, she held the cello and apologized to it (du Pre and du Pre, 1997).

While social responses to artifacts of many types may be common, we have a special relationship to computers because they are unique in their ability to autonomously perform complex cognitive work, and to interact with us in tasks requiring knowledge and judgment. In contrast, traditional machines perform physical work, have no knowledge, and exercise no judgment. Perhaps because of this, people respond to computers as if they were a hybrid between machines and sentient beings even while acknowledging that they "don't really think." People often attribute logic and intentions to computers which they do not really possess.

We provide several examples to illustrate some situations in which people ascribe human-like intent to computer actions. In the first example, users were asked to evaluate Weasel, a computer assistant used in military battle planning, which automatically generates enemy courses of action (ECOA). ECOAs represent hypotheses about possible maneuver plans which enemy forces might follow. The evaluators were shown the relatively small set of rules that Weasel used to construct ECOAs, and the set of ECOAs that it generated for a specific situation (Larson and Hayes, 2005). When asked to explain why Weasel did what it did, evaluators frequently produced complex explanations with many nuances that far exceeded the rules in front of them. "Oh, I see. It (Weasel) is worried about a lateral attack, so it placed these units here to defend this area." Weasel was, in fact, doing nothing of the sort! While the evaluator understood that the COAs were generated by a computer, had read the rules, and had acknowledged the experimenter's explanation that these rules and only these rules were used, he still attributed a more nuanced, human-like logic to the computer. The tendency to attribute human-like powers to computers can be very pronounced.

In the second example, a computer tutor, Adele, monitored medical students as they worked through simulated medical cases (Johnson, 2003). If the student made a mistake, Adele would interrupt and provide feedback about the mistake. If this happened once, the student did not necessarily object, but if it happened multiple times and Adele interrupted and criticized them in the same fashion each time, students came away with the impression that Adele had a "very stern personality and had low regard for the student's work." Adele certainly made no value judgments on the students, and this impression of hostility and rudeness was not intended by Adele's designers.

In the first example, ascribing more detailed and subtle reasoning to the computer than it actually possesses may lead the user to place unwarranted trust in the computer's solutions, also known as overreliance (Parasuraman and Riley, 1997). On the battlefield, this could be life-threatening. In the second example, interpreting the computer tutor as hostile or rude may lead students to disengage from the tutor's lessons, and possibly abandon it altogether; Lewis Johnson's later experiments suggest that students were less motivated and progressed more slowly when their computer tutor was not polite, particularly with difficult problems (Wang et al., 2005).

The point is that a major problem is created by our dual view of computers as chimerical hybrids between machines and intelligent living beings; we consciously design computers as machines, yet we unconsciously respond to them as if they had human-like reasoning

Figure 1.2 Human–Computer Etiquette is a lens through which one may view and explain users' frustrations with computers. (Courtesy of Nancy Johnson).

and intent. When software designers fail to anticipate users' social responses to computers, we may unintentionally violate the rules of engagement for social interaction. Users may become offended, disrupted, angry, or unwilling to use the software (see Figure 1.2). This can lead to reduced performance, errors, and life-threatening situations.

However, it is not necessarily a bad thing to attribute human-like motivations to a machine. It can be very beneficial, as the next example illustrates. When the co-author was soliciting inputs for a project on adaptive information displays for fighter cockpits, an aviator related that every day, when he climbed into the cockpit he asked his aircraft: "How are you going to try to kill me today?" While the aircraft was certainly not designed to kill its own pilot, it was advantageous to the aviator to adopt a mildly antagonistic attitude towards his aircraft, and cast it in the role of an adversary. This helped him to "stay sharp" and anticipate problems while he carried out his work. Specific social relationships with technology can be advantageous and effective—and those relationships need not always be polite or pleasant ones.

1.4 Designing for Appropriate Human–Computer Etiquette

It is becoming more necessary to design computers capable of using appropriate etiquette, especially as computers become capable of increasingly complex cognitive work, in an ever expanding array of roles.

Service-bots (robots that act as personal assistants to people), computer tutors, health coaches, and computer-generated video game characters are all becoming common in our daily lives. Intelligent decision-aiding systems are being used to assist professionals ranging from fighter pilots and nuclear power plant operators, to financial speculators.

But how can one design computers to use etiquette appropriately and effectively? This is relatively uncharted territory. The implicit, contextual, and culturally embedded nature of etiquette makes it challenging for people, let alone computers. The goal of this book is to begin exploration of some of this uncharted territory, and to start sketching in some of its outlines. Some of the questions to be explored include:

- What is human–computer etiquette? What range of behaviors does it entail?
- Are etiquette expectations for computers the same as they are for people?
- Are these expectations modified by the humanness of the computer or robot? For example, do users expect that computers using human voices, faces, or forms to exhibit more human-like interactions?
- How can human–computer etiquette, once designed, be implemented?
- How can we enable computers to gauge the reactions of people, and adjust their behavior accordingly?
- What impact do polite agents have on human emotional response, trust, task performance, user compliance, and willingness to accept a computer application or agent?
- Does the politeness of an application have the potential to influence how we behave online?

The first question, "What is human–computer etiquette?" has many answers. The chapters in this book define etiquette in numerous ways:

- "The culturally embedded expectations for social interactions" (Klein, Chapter 2)
- "Accepted behaviors for a particular type of interaction in a particular group context" (Kaber, Chapter 10)
- "Support (for) social acts that give synergy" which enhance society (Whitworth, Chapter 13)

1.5 Is Human–Computer Etiquette Anything New?

The wide range of definitions and broad set of viewpoints found in the chapters appropriately reflect the complex, rich, and varied nature of the phenomena surrounding human–computer etiquette, and are hence far more useful than a single and precise definition could ever be.

Related to the question "What is human–computer etiquette?" is a subsequent one: "How is human–computer etiquette different from human–computer interaction, social computing, emotional computing, or any number of other research fields and design approaches?" Human–computer etiquette is not distinct from these topics and shares many goals; however, their foci differ. For example, social computing aims to enable and facilitate social interactions between people through or with computers, while human–computer etiquette aims to understand how computers trigger etiquette-based social responses, so that software can be designed to achieve effective interactions.

Human–computer etiquette serves as a lens to view and explain human reactions to computers. It highlights and focuses us on certain aspects of the computer, its role, and the framework of cues and interpretations in a culture, context, or work setting that the computer inevitably enters into. The power of this lens arises from the insight that people respond to computers as if they were hybrids between machines and social beings, even when they insist that they view them as nothing more than mechanical. This lens changes our view of both computers and people, and how to design for them. When viewed through the lens of human–computer etiquette, there is no longer so sharp a distinction between interactions between people and computers, and interactions among people. This view makes clear why software designers can no longer afford to concentrate only on "mechanical" algorithms and logic, but must also consider the social aspects of their software, including whether it will be perceived as kind, trustworthy, rude, or clueless.

The perspectives represented in these chapters provide a framework in which to explore the remaining questions. For software developers, this framework offers another approach with which to improve the users' experiences with, and emotional reactions to, their products. In the current economic climate of intense global competition, product developers are acutely aware that to be competitive, products must not

Figure 1.3 Use of computing and technology in other cultures: female member of Mursi tribe in Ethiopia with rifle and iPod. (Courtesy of iLounge.com).

only function on a technical level, but must also be easy, effective, *and enjoyable* to use. Additionally, we hope these perspectives will provide approaches to inform design of software tools for customers in emerging markets outside Europe and the U.S. where customers may have very different cultural expectations (see Figure 1.3, and Chapter 6). An explicit understanding of human–computer etiquette may allow software designers to better support people's preferred way of working and living in any culture. Finally, by deepening our understanding of the relationships between humans and computers, we may also come to understand more about human relationships.

References

Brown, P. and Levinson, S. C. (1987), *Politeness: Some Universals in Language Usage*. Cambridge University Press, U.K.

Dennett, D. (1987). The Intentional Stance, MIT Press, Cambridge, MA.

du Pre, P. and du Pre, H. (1997), *A Genius in the Family: An Intimate Memoir of Jacqueline Du Pre*, Chatto & Windus.

Gong, L. (2008), The boundary of racial prejudice: Comparing preferences for computer-synthesized black, white and robot characters, *Computers in Human Behavior*, 24(5):2074–2093.

Larson, Capt. A. D. and Hayes, C. C. (2005), An assessment of weasel: A decision support system to assist in military planning, Human Factors Ergonomic Society (HFES) 49th Annual Meeting, September 26–30, 2005, Orlando, FL.

Martin, J. (2004), *Miss Manners' Guide to Excruciatingly Correct Behavior*, W.W. Norton and Company, New York.

Miller, C. (2000), Rules of etiquette, or how a mannerly AUI should comport itself to gain social acceptance and be perceived as gracious and well-behaved in polite society, in Working Notes of the AAAI Spring Symposium Workshop on Adaptive User Interfaces. Stanford, CA; March 20–22.

Miller, C. (Guest Ed.) (2004), Human-computer etiquette: Managing expectations with intelligent agents, *Communications of the Association for Computing Machinery*, 47(4), April 30–34.

Moreno, K., Person, N., Adcock, A., Van Eck, R., Jackson, T., and Marineau, J. (2002), Etiquette and efficacy in animated pedagogical agents: The role of stereotypes, American Association of Artificial Intelligence (AAAI) 2002 Fall Symposium held in Falmouth, MA. AAAI Technical Report FS-02-02. AAAI Press, Menlo Park, CA.

Nass, C., Steuer, J., and Tauber, E. (1994), Computers are social actors, CHI '94, Human Factors in Computing Systems, Boston, MA, pp. 72–77.

Nass, C., Moon, Y., and Green, N. (1997), Are machines gender-neutral? Gender stereotypic responses to computers, *Journal of Applied Psychology*, 27(10): 864–876.

Parasurman, R. and Riley, V. (1997), Humans and automation: Use, misuse, disuse, abuse, *Human Factors*, 39(2): 230–253.

Post, P. (2004), *Emily Post's Etiquette*, 17th edition, Harper Collins e-books.

Wang, N., Johnson, L., Rizzo, P., Shaw, E., and Mayer, R. (2005), Experimental evaluation of polite interaction tactics for pedagogical agents, *Proceedings of the 10th International Conference on Intelligent User Interfaces*, pp. 12–19, San Diego, CA.

Washington, G. (2003), *George Washington's Rules of Civility*, fourth revised Collector's Classic edition, Goose Creek Productions, Virginia Beach, VA.

Weizenbaum, J. (1966), ELIZA—A computer program for the study of natural language communication between man and machine, *Communications of the ACM*, 9(1): 36–45.

PART I

ETIQUETTE AND MULTICULTURAL COLLISIONS

2

As Human–Computer Interactions Go Global

HELEN ALTMAN KLEIN,
KATHERINE LIPPA, AND MEI-HUA LIN

Contents

We have become accustomed to interacting with computers. They give us daily weather forecasts and help us keep in touch with friends and family globally. They deliver references, digital articles, and newspapers and allow virtual teams to work together from opposite sides of the globe. Computers even provide embodied agents to guide us through difficult procedures or teach us to speak Japanese. They allow us to shop for products in distant countries and have ready access to cash on city streets almost everywhere. Human–computer interactions are a part of our lives, mediating communication, providing instruction and support, and serving as gateways for business transactions.

As computer technology spreads around the world, it can act to broaden perspectives as different cultural patterns expand creative options and knowledge while cultivating international understanding and supporting economic, intellectual, and political development. Computer technology introduced to a new nation or region, however, can act as a Trojan horse when the technology conflicts with etiquette, the culturally embedded expectations for social interactions in this nation. Technology may introduce culturally inappropriate interfaces that distort meaning or reduce acceptance. When computers cross national borders, systems designed to support human–computer interactions in one culture may prove to be incompatible with the representations, social expectations, and cognition of other cultures (Shen, Woolley, and Prior 2006). Cultural incompatibility, often invisible, may lead to error, frustration, confusion, conflict, and anger. The rapid expansion of globalization and the unobtrusive nature of important cultural differences have highlighted the need to understand the impact of culture on the way people use computers.

In this chapter, we explore the role of culture in human–computer interactions with the goal of helping international communities take advantage of these powerful tools. Our analysis starts with a description of the Cultural Lens Model as a framework for understanding the ways in which culture shapes how we view and respond to the world. Next, we describe aspects of culture that can affect human–computer interactions. We then look at three domains of computer use: computer-based communication, computers in commerce, and intelligent agent interactions. These illustrate how culture may affect human–computer interactions. Finally, we discuss implications of our analysis and directions for future work.

2.1 How Do Cultures Differ?

While the notion of "culture" can evoke visible expressions such as language, food, and music, culture extends far deeper to embedded social differences in the ways we interact with each other; the values that direct our choices, actions, and plans; and the differences in how we reason, make decisions, and think (Nisbett 2003). Infants around the world start life with virtually identical potentials and tendencies. Distinctive cultural patterns emerge because children typically grow up in a distinct setting where patterns of language, behavior, and thought are shared. During their early years, children learn how to direct their attention, interpret their world, and follow the social rules and roles of their culture. They also adopt ways of making sense of the world, evaluating evidence, and justifying conclusions. The mechanisms of cultural transmission and maintenance are elaborated elsewhere (Berry 1986).

Cultural experiences early in life shape our perception and thoughts. In this respect, culture acts as a multidimensional lens that filters incoming information about the world to provide a consistent view (Klein, Pongonis, and Klein 2000). The *cultural lens* (Klein 2004) includes dimensions suggested by earlier researchers (Hofstede 1980; Kluckhohn and Strodtbeck 1961; Markus and Kitayama 1991; Nisbett 2003). While these dimensions cannot capture the full richness and complexity of cultural groups, they help explain and anticipate the influence of culture in social, professional, and commercial contexts. When people encounter new experiences, these experiences are interpreted through the lens shaped by their early cultural immersion. When people create new objects and systems, cultural dimensions affect how they design these artifacts. While it is possible for new experiences to modify and enhance a person's cultural lens, this takes effort and is not uniformly successful.

Cultural immersion can be both good and troublesome. The good part is that we have a whole set of behavioral scripts, social norms, and cognitive frames that help us function in our environment. We do not have to pay constant attention to acting appropriately. We share a respect for social conventions that allows us to understand, anticipate, and modulate the actions of others within our culture. Our interactions are facilitated by shared social conventions ranging from the etiquette of interpersonal interchanges and the use of a common language to complex aspects of organizational and intellectual activity. Human–human interchanges are supported by common perception, parallel social expectations and patterns, compatible cognition and reasoning, and shared social systems.

The bad part of immersion is that deeply ingrained patterns of behavior and thought can cause confusion, conflict, and paralysis when we leave the confines of our own culture and enter the global community. Cultures define favored logical structures, acceptance of uncertainty, and ways of reasoning. Cultures provide structure for the overall regulation of interactions. When people from different cultures interact, they can be puzzled, dismissive, or even outraged when others ignore social rules, misunderstand ideas, and fail to appreciate "good" judgment in even the simplest of situations.

We assert that cultural differences, powerful forces in human interchanges, are also important in human–computer interactions. Human–computer interactions, like human–human interchanges, are affected by shared cultural patterns and disrupted by conflicting cultural patterns. This is because computer systems designers and programmers embed their own cultural lens in the systems they create. They embed their culture in the system not because of any nefarious motives but because their shared cultural characteristics are simply the most obvious and logical to them. Similarly, computer users call on their own cultural lens as they work with a computer system. If the cultural elements incorporated into the computer and software are inconsistent with the ones the user brings to the interchange, the interaction between computer and user may be conflicted, and usability compromised. In routine computer interactions, be they interacting with distributed teams, searching the Internet, ordering online from a foreign distributor, or sharing a recipe, we can encounter troublesome differences in representation, social expectations, and cognition.

We now describe three kinds of cultural differences among cultural groups that influence human communication and interaction. The first includes culture-specific representational conventions. The second describes the social dimensions of *achievement versus relationship orientation*, *trust*, *power distance*, and *context of communication*. Finally, critical cognitive dimensions include *tolerance for uncertainty*, *analytic versus holistic thinking*, and *hypothetical versus concrete reasoning*.

2.1.1 Culture-Specific Representational Conventions

While differences in language and verbal expression are obvious, cultures also employ representational conventions and symbols to ease the intake and organization of information. These include visual scanning patterns, quantity representation, symbolic representation, and physical features. While Western languages and hence Western computer users orient and read from left to right, this pattern is not universal. For example, Arabic is read from right to left, setting up processing patterns incompatible with most Western systems. The representation of quantity also varies. Americans, for example, expect distances to be measures in inches, feet, yards, and miles; currency, in dollars; and dates ordered as month–day–year. For Europeans, for example, "nine eleven," would normally refer to the ninth day of November, while in the United States, it refers to September 11, 2001. Differences in symbolic meaning are also evident in the symbolic significance of colors and the metaphors used to create icons. In China, red signifies happiness but in the United States it represents danger (Choong, Plocher, and Rau 2005). Finally, ethnic, racial, and gender representations shape our expectations and judgments.

2.1.2 Social Dimensions

Human–human interactions flow most smoothly when each participant in a social interaction follows the same social rules of manners, language, and behaviors. Underlying these surface behaviors, we also expect culturally defined social patterns. These patterns, however, may be altered during computer-mediated interactions. Four culturally defined social patterns illustrate the importance of social dimensions for understanding human interactions.

Achievement versus relationship orientation describes the desired balance between work and social life (Kluckhohn and Strodtbeck 1961). For achievement-oriented cultures, including the United States, the focus at the work place is on the completion of work goals. Work and social relationships are kept relatively separate while objectives are believed achievable through efficient scheduling and task management. For relationship-orientated cultures, personal relationships are a valued part of work. The pace at which tasks are accomplished is seen as less controllable and more relaxed while interactions with fellow workers are more valued.

Trust captures the degree to which people are willing to view themselves as interdependent members of a group. Cultures vary in that in some people define themselves with regard to family or tribe, while in others they look to their age-cohort, workplace, or profession for definition. In-group members, whether they share family, tribal, regional, or professional links, are more likely to be trusted, while the motives and actions of outsiders may receive more scrutiny.

Power distance refers to the relative evenness of the distribution of power (Hofstede 1980). People high in power distance see and accept a world of steep hierarchies and leaders who make decisions with little input from subordinates. People in low power distance cultures, including most Western nations, see a flatter distribution of power, and subordinates are more likely to express ideas and provide input for decision making.

Context of communication. Cultural groups differ in the degree to which messages are communicated explicitly (Hall and Hall 1990). In low-context cultures, such as the United States and Germany, communication is direct and messages are expressed primarily in words. By contrast, in high-context cultures, such as Japan, communication is more indirect, incorporating subtle cues, nonverbal communication, and strategic omissions.

2.1.3 Cognitive Dimensions

People often need to seek information, make sense of complex situations, make decisions, and finally, prepare and implement plans. Cultures differ in their comfort with uncertainty and their use of reasoning strategies. The more compatible information is to a person's culture-linked cognitive processes, the more useful it will be during decision making. The better an argument matches a person's culture-linked reasoning, the more persuasive it will be.

Tolerance for uncertainty is a concept reciprocal to Hofstede's Uncertainty Avoidance (Hofstede 1980), and describes how people react to the unknown. It moderates the evaluation of information and influences planning. Individuals from cultures that are low in tolerance for uncertainty, such as Japan and Korea, find uncertainty stressful. They favor detailed planning and resist change because it can increase uncertainty. People from cultures with a high tolerance for uncertainty, such as the United States, are more comfortable in ambiguous situations, change plans easily, and are more likely to violate rules and procedures that they view as ineffective.

Analytic versus holistic thinking describes how people parse information and environmental stimuli (Choi, Dalal, Kim-Prieto, and Park 2003; Masuda and Nisbett 2001; Nisbett 2003). Analytic thinkers focus on critical objects and features, and tend to organize information based on intrinsic attributes. Holistic thinkers focus on the environmental context and on relationships among elements in the environment. Consistent with this, they typically categorize information based on the relationships between items.

Hypothetical versus concrete reasoning differentiates the degree of abstraction people use in considering future events (Markus and Kitayama 1991). Hypothetical reasoners, typical of the United States and Western Europe, consider future events abstractly. They generate "what if" scenarios and separate potential future outcomes from current and past reality. By contrast, concrete reasoners, characteristic of East Asian cultures, look to the context of the situation and project future outcomes based on past experiences in similar situations.

To illustrate the impact of culture on human–computer interactions, we now explore three domains of computer-based communication, computers in commerce, and intelligent agent interactions.

2.2 Culture in Domains of Human–Computer Interaction

2.2.1 Computers Linking People to People

A man in a cyber café in New Delhi reads an e-mail from his son studying in London. His son is over 4,100 miles away and so he depends on the Internet to keep in touch.

A physician in Japan has to perform a rare and difficult surgery. He arranges a virtual conference with a colleague in Boston. It's crucial he understands all the advice he is given; his patient's life depends on it.

In Internet cafés around the world, people are busy sending e-mail, playing games, reading blogs, and looking for dates. These human–computer interactions allow us to communicate with friends and relatives across national borders and around the world. They provide forums for meeting new people and for developing and maintaining professional partnerships. In 2003, 79% of Americans used the Internet to communicate with family and friends (Fallow 2004). In 2006, 55% of American teenagers used social networking sites (Lenhart and Madden 2007). While computers are playing a growing role in social interactions in many nations, high levels of Internet communication are not yet universal. In particular, sub-Saharan Africa and portions of Asia and South America are still emerging regions. As these social mechanisms continue to migrate around the globe, supportive technologies must adapt to ease access, increase usability, and foster cultural compatibility internationally.

Because the Internet allows growing numbers of people from all over the world to communicate quickly and inexpensively, we can expect to see more situations in which cultural differences lead to misunderstandings in computer-based communication. In relationship-oriented cultures, for example, e-mails, following culture rules of etiquette usually begin with greetings and personal inquiries. Americans, in contrast, are more likely to immediately address the key topic without including a salutation, which may offend someone from a relationship-oriented culture. We spoke with a student from Brazil who described how rude he found the directness of e-mails when he first came to the United States (Mendes 2008). Chinese students similarly reported American e-mail exchanges to seem unfriendly (Xia 2007). Finally, when distributed work teams include members who differ in achievement and relationship orientation, struggles with time allocation and scheduling can compromise the effectiveness of computer-mediated interchanges (see Chapter 3 by Smith). We need culture-sensitive mechanisms to avoid these barriers to cooperative interchanges.

Power distance can also alter the pattern of computer communication by constraining its use to specific purposes and people. A study of a marketing team in South Korea found that team leaders often used e-mails to communicate with the members of their team, but that team members rarely sent e-mails to the leader (Lee 2002). One team member explained this hesitation: "In our Confucius culture, one has to show respect to seniors both in the workplace and at home. I think that e-mail cannot convey signs of respect in an effective manner." Here, users imposed their embedded social rules about interpersonal interchanges on the use of communications technology; communications tools must accommodate differences in context in communication.

We see this in Japan where people avoid expressing their thoughts and feelings directly but rather express them implicitly or communicate them via nonverbal cues. Instant messaging and text messaging systems work for Western communication, which is typically terse and direct, but they present a challenge in Japan. The importance of nonverbal information introduces additional demands. An adaptive solution emerged from this distinctive Japanese characteristic. Traditional

Japanese writing is written using *kanji*, which are ideograms derived from Chinese. There were too many ideograms for use on a keyboard, prompting the adoption of a digital writing form using *kana*, a syllabic alphabet. The adoption of kana provided a large number of symbols for digital writing. Japanese young people used these symbols to create an array of "emoticons"—icons to indicate a complex and nuanced array of mood and facial expression (Kao-moji n.d.). For example:

Smile/laugh	(^^) (^_^) (^–^) (^o^) (^0^) (^O^)
Turn red	(*^^^*) (*^^^*) (#^^#)
Cold sweat/nervousness	(^^) (^^) (^_^) ^^) (^o^) (^_^;)
Embarrassed/surprised	(+_+) (*o*) (?_?) (=_=) (–_–) (°_°)

These are used in instant messaging and text messaging allowing the more indirect, contextualized communication style favored by Japanese users (Sugimoto 2007). Users generated new tools for the expression of emotions during computer interchanges.

Users can sometimes accommodate their culturally embedded expectations by using the expression of emotions and limits on e-mail distribution. In other cases the system itself may need to be adapted to accommodate cultural dimensions. For example, cooperative computer work systems have different requirements in the United States than in Japan. Americans are high in achievement orientation, and so effective teamwork requires a system that supports the interaction and information sharing necessary to get the job done. In line with these proclivities, American systems sometimes provide an online white board and a text-based chat tool focusing on the work being done rather than on relationships to be maintained. By contrast, Heaton (2004) describes a parallel computer cooperative work system developed for use in Japan. In this relationship-oriented culture, personal connections among coworkers are critical. Rather than a traditional white board, the system is based around a "clear board," which creates the impression that collaborators are standing on either side of a glass window. The window displays diagrams and images just like a white board and collaborators can write on it. But, they can also see each other through the board. This allows for more immediacy in the interpersonal interaction and supports the use of nonverbal cues such as gestures and gaze awareness. The provision of the clear board supports all of the technical functions that are typical of a cooperative work system, but the format and support for nonverbal interactions adapts the system to satisfy Japanese values by facilitating interpersonal relationships and high-context communications.

Multiple social and cognitive barriers are captured in three cross-national research efforts that illustrate the multiple social and cognitive difficulties associated with virtual collaborations. First, an educational partnership used Web-based tools to conduct a joint seminar on globalization with United States and South African graduate students (Cogburn and Levinson 2003). They found cultural discrepancies in communication as well as in achievement behaviors as reflected in "work ethic." In a second collaboration, virtual teams with members from Europe, Mexico, and the United States completed a task (Kayworth and Leidner 2000). These teams had intense communication problems. While some were understandably related to language differences, they also reported differences in interpreting e-mails as well as in establishing common ground. Particularly troublesome were achievement behaviors related to meeting deadlines, formality, and planning. Third, in Chapter 3, Smith and colleagues (2010) identify cultural differences in decision making using the C3Fire Microworld's emergency management task.

While technology expands social contacts, it can also inadvertently create misunderstanding, antagonism, and ineffectiveness. Messages are sent that are interpreted as rude and inappropriately cold to people from other groups. We are learning how to translate language meaning; we now need mechanisms that will allow computers to transcend cultural differences, accurately convey subtler nonverbal content, and strengthen interpersonal links.

2.2.2 Computers as Gatekeepers to the International Bazaar

Women at an artisans' cooperative in Kenya make beaded jewelry. Their livelihoods depend on selling their creations via a free trade Web site.

A bank introduces a system to let people from around the European Union automatically pay their monthly bills. The system's acceptance will depend on the trust of depositors in a dozen countries.

Computers are revolutionizing commerce both through the rise of online shopping and through the increasing use of automation as a tool for commercial transactions. A 2007 survey found that over 85% of

Internet users worldwide had purchased something over the Internet, and this percentage is expected to increase. Even in the regions where e-commerce is less frequent—Eastern Europe, Middle East, and Africa—more than 65% of users reported making purchases online (Nielsen Company 2008). These human–computer interactions provide access to commerce worldwide. Social and cognitive expectations, however, influence the willingness of consumers to use computers to buy online, as well as preferences for e-commerce sites.

Hwang, Jung, and Salvendy (2006) compared the expectations and preferences for shopping online as expressed by college students in Turkey, Korea, and the United States. They found that compared to the United States, students in Turkey and Korea wanted significantly more information about products prior to making a selection. In Turkey and Korea, people generally have less tolerance for uncertainty than do people in the United States. This difference may explain their expressed need for more product information. Students in Turkey and Korea also differed in the kind of information they preferred. Korean students wanted comments and opinions from other consumers, as well as technical and price information. They were more reluctant to purchase a product without being able to touch and experience it directly. Compared to Americans and Koreans, Turkish participants showed the most concerns about security. The students were hesitant to entrust financial information to outsiders. This is consistent with Turkish culture that extends less trust to people outside one's own circle of family and friends.

In addition to preferences in the content of e-commerce sites, culture also influences design preferences. Analytic versus holistic reasoning, for example, affects the efficacy of Web site organization. Chinese participants, who are typically holistic thinkers, were more comfortable with and performed better using an online shopping interface in which items were organized thematically. For example kitchenwares—dishwashing liquid, pots, aprons—are grouped together (Rau, Choong, and Salvendy 2004). American participants, in contrast, performed better using a functionally organized group such as cleaning products—dishwashing liquid, soap, and laundry detergent. Holistic thinkers are inclined to focus on the environmental context and on relationships among elements in the environment, whereas analytic thinkers focus on critical objects and features within the environment and organize information functionally. Users experience frustration when online information is perceived as disorganized. An understanding of analytic–holistic differences suggests the need for alternative information structures to meet the needs of different users.

Cultural groups differ in their trust that automated systems will function properly and so differ in their use for commercial activities (Forslin and Kopacek 1992). Computer use also requires a willingness to substitute human–human interaction, with its attendant relationship, for a human–computer interaction. With the increased use of computer-based automated systems for customer transactions, the interaction between automation and culture has taken on a new dimension.

Automatic teller machines (ATMs) exemplify how a system designed to automate a consumer transaction in one culture functioned differently when implemented in another. ATMs were designed in Western nations for rapid, efficient transactions with little emphasis on the human relationships traditionally associated with banking. ATM technology was then exported to India, where speed of the transaction is less important and personal relationship more important. The efficiency of ATM transaction was not generally an incentive for people to adopt the technology. Early adopting Indian users emphasized the social prestige associated with knowing how to use the ATM, rather than its efficiency. They were motivated to adopt technology for the social status it conferred (De Angeli, Athavankar, Joshi, Coventry, and Johnson 2004).

2.2.3 When Computers Masquerade as Humans

A humanitarian emergency calls for the immediate deployment of rescue workers to function effectively and inoffensively in an alien culture. When the task is urgent and pressure is high, there is no place for trial-and-error learning.

Vocational trainers in a refugee camp attempt to accommodate differences in cultural background and preparation. They struggle with large classes that disallow much individual instruction.

Instruction for both rescue workers and school children is best when knowledge and skills are conveyed in ways that are responsive to cultural expectations (Moran and Malott 2004). Computer instruction can use intelligent agents, in the form of embodied agents—human or cartoon faces or full-figure images—to masquerade as responsive trainers to convey needed knowledge and skills. Artificial intelligence allows agents to interact autonomously, providing direct instruction, focus, practice, and guided trouble-shooting. Embodied agents are tools of choice for providing personalized instruction and learning experiences, as in the case of the rescue workers and refugees who need self-paced, responsive instruction.

Embodied agents can interact both verbally and nonverbally by way of gestures, intonation, facial expressions, movement, and turn taking (see Chapter 9 by Bickmore). As we are able to enrich these functions, we increase the believability and utility of the agents. The way in which we do this, however, must reflect cultural expectations.

Embodied agents will be most effective when they assume the cultural representations, social patterns, and cognition patterns of their target users. An instructional program based on the culture of the designer will generally provide a good match with the expectations of users in the designer's culture and will effectively support learning (see Chapter 6 by Rincon et al.). The situation can be different when the embodied agent is provided to users from a different culture. When people interact with agents whose design incorporates a cultural lens different from their own, the interactions are subject to the same challenges as any other cross-cultural interaction. Even a visual representation that includes ethnic identifiers augments effectiveness. Nass, Isbister, and Lee (2000) investigated the effect of ethnic similarity on users' attitudes and choices with Americans and Koreans. When confronted with an agent whose ethnic representation matched their own, both groups of research subjects perceived the agent as attractive and also as more trustworthy. Further, the agent's arguments were seen as more substantive and persuasive. This suggests that the most successful computer mentor will match the ethnicity of the intended learner.

The efficacy of agent design goes beyond ethnic identifiers to other aspects of the culture of anticipated users. Maldonado and Hayes-Roth (2004) created an embodied agent, Kyra, to facilitate an educational program for preteens. Kyra was adapted to interpret and react to incoming information in ways that match American, Brazilian, and Venezuelan expectations. The program incorporated assumptions that specified how agents would interpret and react to incoming information. Modifications were designed to accommodate cultural differences in identity including cultural expectations for women, manner and content of speech adapted to local linguistic patterns, and gestures typical of the culture. These interaction patterns were designed to match those that would be perceived as friendly by each of the target audiences. The accommodations increased the acceptability of the system for preteen users.

Beyond representation, manners, and speech, social dimensions also come into play. In high power distance cultures, students receive information and are expected to master it. Instructors are authorities and direct the learning process. In lower power distance groups, instructors work to support the learning process but the learner is also expected to assume responsibility and provide direction. Achievement versus relationship orientation also contributes to the effectiveness of teaching agents. Westerners, typically achievement oriented, are likely to be comfortable with an embodied agent that presents a professional face and focuses on the task at hand. Learners from relationship-oriented groups would work best with an initial opportunity to "meet" and become acquainted with the agent. This may mean supplying personal information and exchanging social greetings. The nature of communication is also an important social difference for instructional embodied agents. Westerners tend to be direct in their messages. If an answer is incorrect, we may send the learner back to rethink the situation and try again. The learner is likely to assume this to be an opportunity to do it right. In cultures where communication is indirect and honor critical, the straightforward statement of error can be humiliating and discouraging. An embodied agent would need to employ different strategies to allow students to revisit mistakes. Computer instruction will be most effective when the agents adopt instructional styles consistent with the social norms of the learner.

Finally, cognitive characteristics require consideration. In cultural groups characterized by a predominance of concrete reasoning, agents that use this mode of reasoning provide the best initial instruction. A description of a concrete example, such as a specific historic case, would make instructions more relevant; while a hypothetical argument would reduce the perceived credibility, believability, and trustworthiness of the agent, making later learning more difficult. In groups characterized by holistic reasoning, categorization using relationships and situation-sensitive options consistent with cultural preferences would be most effective. Overall, the computer can be most effective when it masquerades as a member of the learner's cultural group adopting strategies and interaction patterns consistent with cultural preferences.

2.3 Implications

2.3.1 Users

As users, we must start by realizing that cyberspace is a world without national borders. We each arrive in this world with a distinct cultural lens and perspective. Power distance moderates the tone and direction of interchanges and suggests how best to carry out interactions. In high power distance groups respect and formal terms of address are important and the distribution of information may be limited. Achievement versus relationship orientation shapes the type and quantity of personal information shared. Too much or too little self-disclosure may be considered rude. When formulating messages, we may need to acknowledge the subtleties of the context of communications to avoid misinterpretation and insult. Differences in analytic versus holistic thinking and hypothetical versus concrete reasoning can affect whether the information we share is understood or misunderstood, and how it is evaluated.

Users of all types benefit from understanding cultural differences. Teenagers can make friends around the world. Merchants can optimize the virtual presentation of their wares and the likelihood that virtual customers will develop sufficient trust to make purchases. Entrepreneurs can form new partnerships. And, people from many different disciplines—artists, doctors, scientists, philosophers—can collaborate and share information with a broader community of colleagues.

2.3.2 Designers

As designers use their expertise in hardware and software to provide intelligent functionality, they can easily and unconsciously assume their own culture's lens (Laplante, Hoffman, and Klein 2007). Doing so may confuse or alienate potential users from other cultures. To avoid this "designer-centered design" error, they will have to make accommodations for the culture-related behavioral, social, and cognitive differences that influence human–computer interactions. Designers must first identify who their potential users are, what they need the system to do, and how the system and users relate to each other. They must understand cultural differences in how activities are approached and how people collaborate on tasks. These include expectations about functions the software should support and preferences for specific system features such as trouble-shooting assistance. Cultural characteristics important for the target domain and expectations about social context of work are particularly important. How tasks are divided, what information should be available to whom, and what interactions are likely to be necessary or desirable to accomplish the task are important questions. These decisions depend on the categories expected, customary organization, relative importance of achievement and relationships, and power distance.

Designers of artificially intelligent and embodied agents may be most successful when they accommodate cultural differences in their designs by adopting the avatar concept from the gaming world. Users can select the physical and nonphysical characteristics of the agent with which they wish to interact. The human–computer interaction will then assume more relevance for the user.

Documentation, training, and support can also help to bridge the gap between the system's capacities, characteristics, and demands on one hand, and the successful acceptance and use of the system on the other. Supporting materials help users access the system and learn to exploit its capabilities. People from different cultures vary in how they approach learning about new systems and in how they anticipate and manage problems. While culture-related guidelines for computer system documentation, training, and support are scarce, the cultural training literature, together with the literature on cultural differences, can clarify decisions about the organization, provide requirements for documentation, inform the development of training material and instructional approaches, and suggest appropriate support services.

2.3.3 Future Directions

We need to learn more about how interface design can accommodate cultural differences in language, symbols, and representations.

Systems need to support expected communication paths and allow for the control of information in culturally endorsed ways. Power distance, trust, and context of communication can affect the choice of who we should communicate with and the appropriate method of communication. Systems designed for international use must provide not only text-based communication but also support for indirect communications, characteristic of many new user nations. Because cultural groups vary in their expectations about how personal relationships and work interactions are best balanced, human–computer interactions must accommodate both expectations.

The internationalization of human–computer interaction will be a fruitful target for future work. Armed with knowledge of cultural differences, we start to identify barriers to effective interactions. As we can identify these barriers, we can begin to address the attendant challenges. Our interactions via computers are taking us to new places with unexpected challenges. We will need to remain vigilant for differences in emerging national groups. We are beginning a long and interesting journey, and the payoff can be substantial.

References

Berry, J.W. 1986. The comparative study of cognitive abilities: A summary. In *Human Assessment: Cognition and Motivation,* ed. S.E. Newstead, S.H. Irvine, and P.L. Dann. The Netherlands: Martinus Nijhohh.

Choi, I., R. Dalal, C. Kim-Prieto, and H. Park. 2003. Culture and judgement of causal relevance. *Journal of Personality and Social Psychology* 84, no. 1: 46–59.

Choong, Y-Y., T. Plocher, and P-L.P. Rau. 2005. Cross-cultural Web design. In *Handbook of Human Factors in Web Design,* ed. R.W. Proctor and K-P.L. Vu, 284–300. Mahwah, NJ: Lawrence Erlbaum Associates.

Cogburn, D.L. and N.S. Levinson. 2003. US-Africa virtual collaboration in globalization studies: Success factors for complex, cross-national learning teams. *International Studies Perspectives* 4: 34–51.

De Angeli, A., U. Athavankar, A. Joshi, L. Coventry, and G.I. Johnson. 2004. Introducing ATMs in India: A contextual inquiry. *Interacting with Computers* 16, no. 1: 29–44.

Fallow, Deborah. 2004. The internet and daily life: Many Americans use the Internet in everyday activities, but traditional offline habits still dominate, Pew Internet and American Life Project. www.pewinternet.org/ (accessed February 22, 2008).

Forslin, J. and P. Kopacek. 1992. *Cultural Aspects of Automation*. New York: Springer-Verlag.

Hall, E.T. and M.R. Hall. 1990. *Understanding Cultural Differences.* Maine: Intercultural Press Inc.

Heaton, L. 2004. Designing technology, designing culture. In *Agent Culture: Human-Agent Interaction in a Multicultural World,* ed. S. Payr and R. Trappl, 21–44. Mahwah, NJ: Lawrence Erlbaum Associates.

Hofstede, G. 1980. *Culture's Consequences: International Differences in Work-Related Values.* California: Sage.

Hwang, W., H.S. Jung, and G. Salvendy. 2006. Internationalisation of e-commerce: A comparison of online shopping preferences among Korean, Turkish and US populations. *Behaviour and Information Technology* 25, no. 1: 3–18.

Kao-moji (Japanese Emoticons). (n.d.) http://www2.tokai.or.jp/yuki/kaomoji/index.htm (accessed February 27, 2008).

Kayworth, T. and D. Leidner. 2000. The global virtual manager: A prescription for success. *European Management Journal* 18, no. 2: 183–194.

Klein, H.A. 2004. Cognition in natural settings: The cultural lens model. In *Cultural Ergonomics: Advances in Human Performance and Cognitive Engineering,* ed. M. Kaplan, 249–280. Oxford: Elsevier.

Klein, H.A., A. Pongonis, and G. Klein. 2000. Cultural barriers to multinational C2 decision making. Paper presented at the *Proceedings of the 2000 Command and Control Research and Technology Symposium (CD-Rom)*, Montgomery, CA.

Kluckhohn, Florence Rockwood and Fred L. Strodtbeck. 1961. *Variations in Value Orientations*. Illinois: Row, Peterson.

Laplante, P., R.R. Hoffman, and G. Klein. 2007. Anti patterns in the creation of intelligent systems. *IEEE Intelligent Systems.* January–February 2007, 91–95.

Lee, O. 2002. Cultural differences in e-mail use of virtual teams: A critical social theory perspective. *Cyber Psychology and Behavior* 5, no. 3: 227–232.

Lenhart, A. and M. Madden. 2007. Social networking websites and teens: An overview, Pew Internet Project. www.pewinternet.org/ (accessed February 22, 2008).

Maldonado, H. and B. Hayes-Roth. 2004. Towards cross-cultural believability in character design. In *Agent Culture: Human-Agent Interaction in a Multicultural World,* ed. S. Payr and R. Trappl, 143–175. New Jersey: Lawrence Erlbaum Associates.

Markus, H.R. and S. Kitayama. 1991. Culture and the self: Implications for cognition, emotion, and motivation. *Psychological Review* 98, no. 2: 224–253.

Masuda, T. and R.E. Nisbett. 2001. Attending holistically versus analytically: Comparing the context sensitivity of Japanese and Americans. *Journal of Personality and Social Psychology* 81, no. 5: 922–934.

Mendes, P. Interview by Helen Altman Klein. Interview format. Wright State University, February 18, 2008.

Moran, D.J. and R.W. Malott. 2004. *Evidence-Based Educational Methods.* San Diego, CA: Elsevier Academic Press.

Nass, C., K. Isbister, and E.-J. Lee. 2000. Truth is beauty: Researching embodied conversational agents. In *Embedded Conversational Agents,* ed. S. Prevost, J. Cassell, J. Sullivan, and E. Churchill, 374–402. Massachusetts: MIT Press.

Nisbett, R.E. 2003. *The Geography of Thought: How Asians and Westerners Think Differently and Why*. New York: The Free Press.

Rau, P-L.P., Y-Y. Choong, and G. Salvendy. 2004. A cross-cultural study on knowledge representation and structure in human computer interfaces. *International Journal of Industrial Ergonomics* 34, no. 2: 117–129.

Shen, S-T., M. Woolley, and S. Prior. 2006. Towards culture-centered design. *Interacting with Computers* 18, no. 4: 820–852.

Smith, K., I. Lindgren, and R. Granlund, R. 2010. Exploring cultural differences in team collaboration. In *Human-Computer Etiquette*, ed. C. Hayes and C. Miller, pp–pp. Boca Raton, Florida: Taylor & Francis Group.

Sugimoto, T. 2007. Non-existence of systematic education on computerized writing in Japanese schools. *Computers and Composition* 24, 317–328.

The Nielsen Company. 2008. Over 875 million consumers have shopped online—The number of internet shoppers is up 40% in two years. Press release, New York.

Xia, Y. 2007. Intercultural computer-mediated communication between Chinese and U.S. college students. In *Linguistic and Cultural Online Communication Issues in the Global Age,* ed. K. St. Amant, 63–77. Pennsylvania: Information Science Reference.

3

Etiquette to Bridge Cultural Faultlines

Cultural Faultlines in Multinational Teams: Potential for Unintended Rudeness

KIP SMITH, REGO GRANLUND, AND IDA LINDGREN

Contents

3.1 Introduction

The cultural faultline model provides an accessible metaphor that explains and predicts where and why some multicultural teams cohere while others collapse into a morass of misunderstanding. In this chapter, we outline the model, present our microworld methodology for eliciting cultural differences in teamwork, discuss four prototypical dimensions of cultural difference in teamwork that emerged in our laboratory, and prescribe how designers of technology might employ etiquette to bridge these and other cultural differences that can be expected to arise in the course of technology-mediated teamwork.

3.2 Etiquette to Bridge Cultural Faultlines

The basic premise of this chapter is that the impact of culture on technology-mediated interactions among team members parallels its direct communication between people. Further, we can expect situations in which the match or mismatch between diverse cultures and technological support for their accustomed modes of communication will generate confusion or discord, rather than understanding (see Chapter 1; Klein, 2005). To minimize the likelihood of disadvantageous outcomes, the design of technology must incorporate etiquette that spans cultural faultlines and facilitates cohesion in multicultural teams. To succeed at that task, designers need to understand how people from different cultures spontaneously elect to approach their task and collaborate on it. The understanding designers seek is culture and task-specific. How do people from two cultures approach the same task? How do these approaches differ? Are these cultural differences likely to lead to paralysis or to support team cohesion? How can etiquette help a multicultural team working on a common task to bridge the cultural faultlines that threaten the team's cohesion?

This chapter seeks to answer those questions for four diverse cultures and a singularly time-critical task. We begin by presenting a model that explains and predicts the impact of cultural differences on human–human interactions and, by extension, their impact on technology-mediated interaction. The model draws upon and extends the "group faultline" model proposed by Lau and Murnighan (1998; 2005). Lau and Murnighan introduced the group faultline model to explain the impact of demographic diversity on the effectiveness of work groups. We propose that the group faultline model can and should be extended to encompass dimensions of cultural diversity, as well as demographic characteristics.

The second section discusses our microworld methodology for uncovering cultural differences that spontaneously emerge in the course of teamwork. Our approach is to simulate a time-critical task using the C3Fire microworld (Granlund, 2002; 2003), to conduct dynamic laboratory experiments with culturally homogeneous teams, and to record the actions the teams took and the decisions they made. We work with culturally homogeneous teams in order to identify that culture's norms for teamwork. We argue that microworlds can and should be used more generally to reveal cultural norms for task performance and collaboration.

In the third section, we review four prototypical dimensions of cultural diversity that spontaneously emerged in our laboratory. We present these findings with a caveat: the dimensions of diversity that we identified are not intended as definitive characterizations of specific national groups; rather, they are discussed as exemplars of the variety of faultlines that are likely to appear whenever multinational teams are formed. We conclude by drawing upon these findings to prescribe how etiquette could be used to bridge the faultlines formed by the cultural differences we found.

3.3 The Cultural Faultline Model

Culture can be seen as a group's shared/collective attitudes, beliefs, behavioral norms, and basic assumptions and values that provide a lens for perceiving, believing, evaluating, communicating, and acting (Klein, 2004; Triandis, 1996). A culture is shared by those with a common language within a specific historic period and a contiguous geographic location. It is passed down from one generation to the next. This heritage influences how people think, speak, and act, and cannot easily be ignored (Kim and Markus, 1999; P. B. Smith and Bond, 1999). For succinctness, we adopt Smith and Bond's (p. 39) definition of culture: "A culture is a relatively organized system of shared meanings." This definition is sufficiently broad to apply to professional cultures, regional cultures, and national cultures and to differentiate among cultures of each type. The focus of our empirical research has been the diversity of norms for teamwork held by four different national cultures.

3.3.1 Demographic Faultlines

The term "diversity" typically refers to the degree to which members of a group have different demographic attributes such as gender or ethnicity. For good or bad, categorizations based on salient personal attributes provide the initial impressions on which groups begin to interact and cooperate. The research on diversity in work groups has not produced consistent results (Thatcher, Jehn, and Zanu, 2003). Many studies show that diversity increases creativity and improves performance. Many others show that diversity spawns conflict and undermines teamwork. Lau and Murnighan (1998) point to one reason for this inconsistency: diversity research has traditionally assessed the impact of only one demographic characteristic (such as gender or ethnicity) at a time. They argue that any analysis of diversity must go beyond the consideration of single characteristics in isolation and investigate the effects of multiple characteristics and their interrelationships. To address this methodological lapse, Lau and Murnighan introduced the "group faultline" model to explain the impact of demographic diversity on the effectiveness of work groups. Their article has spawned a growing literature on group faultlines (e.g., Lau and Murnighan, 2005; Molleman, 2005; Lindgren and Smith, 2006; K. Smith, Lindgren, and Granlund, 2008; Thatcher et al., 2003).

Group faultlines are hypothetical dividing lines that may split a diverse group into subgroups based on several characteristics simultaneously (e.g., nationality and gender) and their alignment. As an illustration, consider the following two groups. Group A is composed of two Swedish women and two Bosnian men. Group B is composed of one Swedish woman, one Swedish man, one Bosnian woman, and one Bosnian man. In both groups there are two nationalities and two genders. In Group A, differences in both characteristics align. In Group B, they do not. The group faultline model maintains that the alignment of multiple characteristics makes Group A more likely to split into subgroups than Group B. By analogy to geological faultlines, there is a faultline between the two pairs in Group A that has the potential to generate friction as subgroups on either side attempt to move in different directions. Faultlines pose a barrier to team cohesion. The faultline metaphor provides a vocabulary for articulating the sources of group dynamics and a mechanism for predicting the impact of diversity.

Lau and Murnighan (1998) differentiated between "strong" and "weak" faultlines, with strong faultlines being those that form when multiple attributes are aligned in a manner that defines clear subgroups. This terminology is at odds with the geologic metaphor. All geologic faultlines are (or once were) planes of weakness. In keeping with the geologic metaphor, we differentiate between "long" and "short" faultlines. The length of a group faultline depends on three compositional factors: (1) the number of individual characteristics apparent to group members, (2) their alignment, and, as a consequence, (3) the number of potentially homogeneous subgroups.

When the group is new, faultlines are most likely to form based on demographic attributes. As members interact, other personal attributes such as personality, values, and skills become increasingly influential and may in turn lead to the development of new faultlines (Lau and Murnighan, 2005). Depending on the similarity and salience of the members' characteristics, a group may have many potential faultlines, each of which has the potential to activate. Active faultlines increase the potential for the group to split into subgroups composed of individuals with similar (aligned) characteristics. A team that splits apart is likely to be ineffective.

3.3.2 Cultural Faultlines

We have proposed that the faultline metaphor can and should be extended to encompass dimensions of cultural diversity as well as demographic characteristics (K. Smith et al., 2008). For example, consider a multinational team composed of two Swedish men and two men from a culture that more readily accepts centralized, authoritarian decision making. The difference in their norms for the process of decision making would align with their demographic and linguistic diversity. If the group proceeds without sufficient coordination, this alignment of cultural and demographic sources of diversity may lead to activation of a group faultline. The superposition of diverse cultural norms on the more obvious demographic differences might be sufficiently salient to activate the faultline and destroy any semblance of group cohesion. We propose that this "cultural faultline" model is a natural extension of Lau and Murnighan's group faultline model.

3.4 Methodology

Our approach to investigating and capturing teamwork in a dynamic and complex work situation is to use the C3Fire microworld (Granlund, 2002; 2003). C3Fire simulates an emergency services management task and elicits distributed decision making from a group of decision makers.

Microworlds are distributed, computer-based simulated environments that realistically capture much of the complex, opaque, and dynamic nature of real-world problems to groups of participants (Brehmer, 2005; Brehmer and Dörner, 1993; Johansson, Persson, Granlund, and Matts, 2003). They capture complexity by posing multiple and often conflicting goals that force decision makers to consider disparate courses of action simultaneously. The problems microworlds pose are dynamic in the sense that the decision makers have to consider the interdependencies of their actions and the unfolding of events at multiple time-scales. Like many real-world problems, microworlds may contain environmental forces that are largely invisible to the participants, generating cascades of unforeseen consequences.

The experimenter has control over much, but not all, of the complexity, opacity, and dynamics of the microworld. The experimenter can set the stage and pose the problem but cannot control how the participants interact with the simulated environment, or how that interaction unfolds over time. Thus, experimental trials that start at the same point within a microworld may or may not progress in the same way. The evolution of events depends on the actions and decisions taken by the participants. This trade-off between strict experimental control and the realism of contextual variability makes microworld experiments more complex, challenging, and realistic than traditional laboratory studies. It also makes them more controllable and easier to analyze than observational field studies. In this way, microworlds bridge the gap between the confines of the traditional laboratory experiment and the "deep blue sea" of field research (Brehmer and Dörner, 1993). By capturing the drivers of real world problem solving, microworlds engage participants. Participants take well-designed microworlds seriously and become so engaged that their actions become completely natural and, accordingly, valuable to the researcher (Gray, 2002, Strohschneider and Güss, 1999).

Our unit of analysis is the team. By investigating how teams from four different national groups approached the task posed by the C3Fire microworld, we were able to identify divergent cultural norms for team collaboration.

3.4.1 Participants

Since it is difficult to know exactly how to distinguish one culture from another based on something other than nationality, we used nationality as our proxy for cultural heritage. Using nationality as a "definition" of culture is widely recognized to be a convenient solution at best (e.g., Hofstede, 1980; Schwartz, 1992; P. B. Smith, Bond, and Kağitçib, 2006) that has been roundly and appropriately criticized (e.g., Duranti, 1997; Hofstede and Hofstede, 2005; Matsumoto, 2003). We are aware of the difficulties in doing so, but as we wish to identify cultural diversity in norms for group behavior and as we have to work within our means, nationality is our best option. We do not claim that the results from our teams can be generalized to all individuals in their nations of origin. Rather, we assume that the differences in their actions in response to identical situations can be explained by their cultural heritage.

A total of 114 participants (6 women and 108 men; mean age 25 years) who identified themselves as either Swedish, Bosnian, Indian, or Pakistani participated in our experiments. We will use the abbreviations: S, B, I, P to represent each group, respectively. We avoided potential demographic confounds by keeping the demographic characteristics of our participants as homogenous as possible. In each experimental group all participants (1) were the same sex, (2) were approximately the same age, (3) had approximately the same level of education, and (4) came from the same country. The age and education matches applied across national groups as well. The matched sampling facilitates comparison across the national groups.

By recruiting only men, we had hoped to eliminate gender as an unaccounted contextual variable. However, when recruiting Bosnian participants, a group of women reported interest in participating. We therefore had one group of six Bosnian women in an all-women session.

The Swedish participants were native Swedes. Most were students at Linköping University. The Swedish participants who were not students had a university degree. The Bosnian participants were born and, to some extent, raised in Bosnia. Half of the Bosnian participants were students at the University of Skövde. The other half worked for local industry. All Indian and Pakistani participants were graduate students at the universities in Linköping and Skövde. The Swedish and Bosnian participants had similar educational backgrounds. The Indians and Pakistanis, however, were slightly more educated. Several of the Indian and Pakistani participants were in Sweden to pursue a second master's degree. In their response to a questionnaire, all participants indicated they used computers for work or entertainment or both. Their computer literacy included word processing and chat programs.

All participants signed an informed consent form. The rules and regulations of the Human Subjects Committee of Linköping University were adhered to at all times. Each participant was promised monetary compensation of 500 Swedish kronor (approximately $70) for completing approximately eight hours of experimentation. All participants completed the study and received their compensation.

In this chapter we use the term "team" to refer to the ad hoc groups of participants who worked together during our experimental sessions. We are fully aware that these groups are not true teams but need to use the "word" team to distinguish between the small groups of participants and the larger national groups. We reserve the word "group" for the national groups.

3.4.2 Apparatus

We used the C3Fire microworld (Granlund, 2002, 2003; Johansson et al., 2003; Rigas, Carling, Breh, 2002) to present an emergency management task to teams of three or four participants. C3Fire is a computer-based platform that uses a server–client architecture to provide an environment for the controlled study of collaborative decision making in a dynamic environment (Lindgren and Smith, 2006). Each participant works at his or her own client PC. Every keystroke and every event in an experimental trial generate time-stamped data that are logged by the C3Fire server.

3.4.3 Task

The emergency presented by C3Fire is a forest fire. The team's task was to manage or suppress the fire. The interface contains a map, an e-mail facility, and information about the status of firefighting equipment. All participants saw the same interface and the same map representation of the simulated world, and were presented with the same complete and accurate information. The map is divided into a grid of squares. Each square is color-coded to represent a combination of terrain and vegetation. Some cells also contain icons representing houses, schools, water stations, and fuel stations. The speeds with which the fire burns and spreads are functions of vegetation, terrain, the presence of buildings, wind direction, and wind speed.

To manage the fire, the team had access to six fire trucks, three water trucks, and three fuel trucks. Every member of the team could direct all 12 trucks. A participant dispatched the trucks by using the computer mouse to direct trucks to move to cells in the map grid. A fire truck that stands on a cell that is on fire automatically attempts to suppress the fire. To do so, it needs water. To move to the fire, it needs fuel. Water trucks have large water tanks and can provide the fire trucks with the water they need. Similarly, fuel trucks can supply both fire and water trucks with fuel. The water and fuel trucks can be refilled at water and fuel stations. Trucks are constrained by pre-set limits on the rates with which they move and act (e.g., fight fire, fill with water).

Interdependencies among team members arise whenever different types of truck are assigned to different participants. For example, the locations and activities of water trucks and fuel trucks constrain the actions of the fire trucks. If different participants have control over these different resources, their actions are mutually constraining. This provides ample opportunity for intra-group conflict.

The participants were asked to communicate only via the C3Fire e-mail system. The interface contains separate windows for sending and receiving messages. The sender of a message was able to specify a particular recipient or send the message to all other participants. We, as experimenters, did not constrain the team's communication in any way other than by asking that all messages be written in either Swedish or English. Similarly, we did not assign roles or establish an organizational structure for truck control. As a result, each team member could (1) communicate with all other members, (2) command all trucks and, (3) override commands made by other participants. In short, all organizational and communication structure was left to the teams.

3.4.4 Procedure

Volunteers scheduled to report to the laboratory in culturally homogeneous (and same-gender) groups of eight. On several occasions only six or seven volunteers actually showed up. In the laboratory, after reading the instructions to subjects and signing the informed consent forms, the participants completed a series of self-paced, individual training trials that taught them how to dispatch and refuel the trucks and use the e-mail facility. After everyone reported feeling comfortable with C3Fire, they were assigned to teams of four and completed a pair of group-training trials.

After the training trials and a short break, the participants were randomly assigned to two teams and to different server computers. The two teams worked in parallel to manage two different simulated forest fires. This arrangement made it possible to gather data on two teams simultaneously. The teams performed eight cycles of two activities. The first activity was a C3Fire experimental trial. Participants sat at separate client computers and were linked together by C3Fire. Each trial lasted until the fire had been put out or 20 minutes had passed. After each trial, the experimenter led the teams in an after-action review during which they engaged in open-ended conversations about their play. Most teams discussed how responsibilities were to be allocated in the next trial and debated alternative strategies for managing the fire.

We created eight different experimental scenarios by manipulating three factors: map, map rotation, and initial fire size. Two different maps with differing configurations of forests and houses, etc., form the foundation for the eight scenarios. Each map was presented four times, at four different rotations (0°, 90°, 180°, and 270°), to make the maps appear different. As no participant mentioned that the same map had been used more than once, this manipulation appears to have been effective. Initial fire size refers to the size of the fire, in squares, at the beginning of the scenario. This was manipulated at two levels (number of squares: 2 × 2 and 3 × 3). The larger the fire, the greater the challenge.

3.4.5 Data

For the duration of an experimental trial, the C3Fire system monitors the status of the fire and trucks (e.g., water and fuel levels), when and where team members dispatch trucks, and all e-mail communication. It creates a log with all events in the simulation (e.g., when and where the fire spreads or a truck runs out of fuel), each participant's commands to the trucks, and all of the team's e-mail communication. In this chapter, we restrict our discussion to three of the many data sets captured by the C3Fire system: where participants dispatched trucks, who was commanding each truck, and how frequently, and what e-mail communications were made. These data inform analysis of the goals that the team pursued, their allocation of roles and responsibilities across team members, their use of feedback, and the role of etiquette in bridging cultural faultlines.

3.5 Cultural Differences in Teamwork

We begin this section by presenting data on the teams' performance during the C3Fire trials. We then turn to task allocation structures and e-mail communication. For each topic, we discuss how the national groups align. To foreshadow the findings, we present evidence of four alignment patterns observed in the experiments. We use the notation (S // BP // I) for the first alignment indicating that Bosnians and Pakistanis share a norm for teamwork that differs from the Swedish norm and from the Indian norm, and the Swedish norm differs from the Indian norm. We use the notation (S // P // BI) for the second alignment: Bosnians and Indians share a norm that differs from the Swedish and Pakistani norms. Similarly, the notation (S // BPI) indicates that the Swedish norm differs from that shared by the other three national groups. Finally, notation (SB // PI) represents the fourth alignment: Swedish and Bosnians share a norm that differs from those shared by Pakistanis and Indians.

3.5.1 Goal Setting

The instructions to subjects informed the teams that their task was to "manage the fire." No additional guidance was provided. The teams were free to establish goals and dispatch their trucks as they chose. Through their interaction with C3Fire, the teams revealed their norms for goal setting.

Figure 3.1 illustrates the cultural differences revealed by the experiments in national norms for goal setting. The three maps show all the locations where fire trucks were dispatched during a typical trial. The grids represent the C3Fire map. The small open squares show all locations to which fire trucks were dispatched at some point during the trial. Because most trucks were dispatched to several cells over the course of a scenario, each truck is represented several times. The large rectangles that surround most of the fire trucks are defined by the ±95% confidence intervals in X and Y for fire truck locations. The small dots show the locations of houses, schools, water stations, and fuel stations.

Figure 3.1a shows the locations of trucks during a typical trial with a Swedish team. The trucks are concentrated directly on the fire (not shown). This rather densely packed set of truck placements is consistent with the goal to attack the fire head-on and suppress it. Figure 3.1b shows a Pakistani trial that is typical of both Bosnian and Pakistani teams. This pattern suggests the goal to contain the fire by forming fire breaks. Figure 3.1c shows where fire trucks were dispatched in a typical Indian trial. The collocation of fire trucks and habitations supports the goal of protecting people and their homes.

To quantify differences in performance related to goal setting, we used the 95% CI rectangles to calculate a metric of truck density. (Density = area of the rectangle/number of truck locations). The density metric is bounded by 0 and 1. Across all 234 trials, values range from .07 to .61 with a median of .25. A one-way ANOVA indicates a significant difference across national groups, $F(3, 230) = 6.19$, $MSE = 0.048$, $p < .001$, power > .92. The Tukey HSD procedure indicates that (a) the density of fire trucks dispatched by the Indian teams was

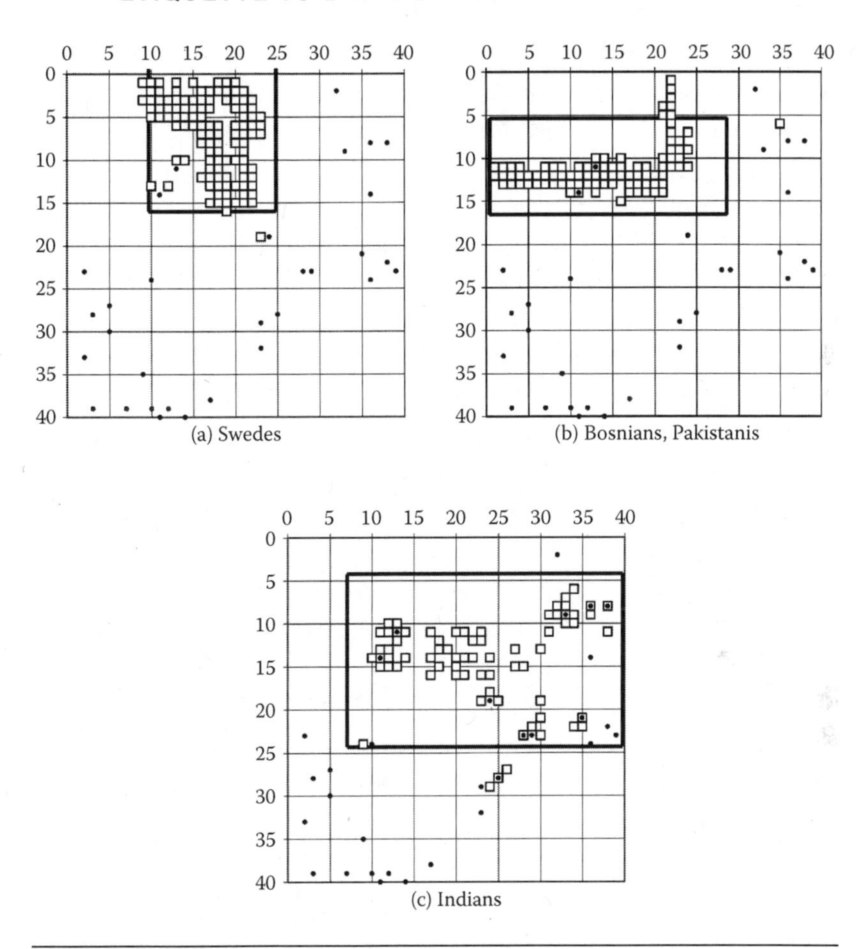

Figure 3.1 Maps showing typical distributions of fire trucks during a C3Fire trial by national group. (a) Most Swedish teams attacked the fire. (b) Most Indian teams protected houses and school (shown as dots). (c) Most Bosnian and Pakistani teams built fire breaks.

significantly less than it was for every other national group, (b) the density of trucks by the Bosnian and Pakistani teams was statistically identical, and (c) the density of trucks by the Swedish teams was greater than the Bosnian and Pakistani teams, but the difference was not significant at the .05 level.

The density metric quantifies the observation that the four national groups pursued three different goals with respect to fire truck placement when given the same mission. The Swedes pursued the goal of suppressing the fire by concentrating the trucks in a dense pack. The Bosnians and Pakistanis managed the fire by spreading the trucks and creating fire breaks. Their containment strategy reduced the average density of fire trucks. The Indians dispatched trucks to the vicinity of houses. Because the houses were widely scattered across the map, their strategy scattered the trucks and produced relatively low densities. This strategy left much of the C3Fire world in ashes. The Indians "sacrificed" much of the vegetation to save the houses. The other national groups sacrificed a few houses to suppress or contain the fire.

We use the notation (S // BP // I) to represent a pair of cultural faultlines between three alignments of cultural norms: (a) the alignment of goals set by our Bosnian and Pakistani teams, (b) the disparity between that goal and the Swedish and Indian goals, and (c) the disparity between the Swedish and Indian goals.

3.5.2 Task Allocation

We turn now from a measure of team performance to a measure of collaboration: the distribution of trucks across the members of a team. Truck assignment in C3Fire reflects the allocation of roles and responsibilities within the team. By capturing who issued commands to which truck, we can assess cultural differences in norms for team organization.

Figure 3.2 presents an example of our matrix representation of the relative frequency of commands sent by team members (A, B, C, and D) to the 12 trucks (F1, F2, F3, etc.). Rows represent the team members; columns represent trucks. A fully black cell represents the highest percentage of commands sent to a truck during the trial. At the other extreme, a purely white cell indicates that no commands were sent to that truck by that participant. Intermediate tones of grey represent intermediate percentages of commands in a linear mapping. Two cells that are equally dark represent equal frequencies of commands. In Figure 3.2, we can see that participant A issued no commands; participant B sent commands only to gas trucks (G10-12) and participant C only to water trucks (W7-9). In contrast, participant D sent commands to almost all trucks, but concentrated on the fire trucks (F1-6). This distribution suggests that the team largely adhered to a relatively strict partitioning of roles and responsibilities.

We used matrices from the 234 trials to develop the taxonomy of task allocation structures shown in Table 3.1. Strict rules for seven different categories of task allocation were set and written down. The left-hand column of Table 3.1 lists the rules; the right-hand column

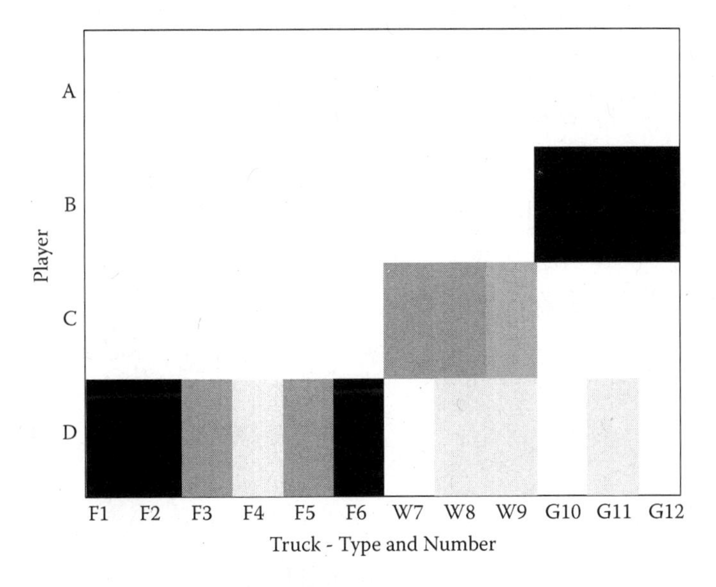

Figure 3.2 An example matrix of team members and trucks showing the relative frequency of commands to trucks: Fire trucks 1–6, water trucks 7–9, and fuel trucks 10–12. Cell darkness increases with the frequency of commands.

shows illustrative examples of the categories. Two coders were used to ensure the coding was conducted according to the categories.

The two "Partitioned" categories are task allocation structures in which each team member commanded three trucks. Teams that adopted these approaches revealed a norm for partitioning tasks equitably among team members. In contrast, the "Assistant" and "Coordinator" categories represent formal, hierarchic allocations of responsibilities. A nominal leader monitored the C3Fire interface and sent e-mails to the team members with recommendations for what needs to be done. The two "Shared" categories represent truly cooperative approaches to the task. Finally, the "Open" structure contains matrices in which a visible task allocation structure is essentially absent.

Table 3.2 summarizes the distribution of allocation structures across the national groups. The distribution of trucks differs significantly across the national groups, $\chi^2(18, N = 234) = 119.4, p < .001$. The Assistant strategy was used by all three groups, but mostly by Swedish teams. The Coordinator strategy was used infrequently. Our Swedish teams most frequently opted for partitioning based on convenience. Our Bosnian and Indian teams often opted for partitioning based on preference. The Pakistani teams used both of the partitioned

Table 3.1 The Taxonomy of Organizational Structures

<table>
<tr><th>CATEGORY</th><th>DESCRIPTION</th><th>EXAMPLE</th></tr>
<tr><td>Partitioned according to "convenience"</td><td>The participants command three trucks each. The partition is based on participant name and truck number. Participant A: trucks 1–3; B: trucks 4–6; C: trucks 7–9, D: trucks 10–12. (In teams with three participants: A: fire trucks 1–6; B: water trucks 7–9; C: gas trucks 10–12.)</td><td>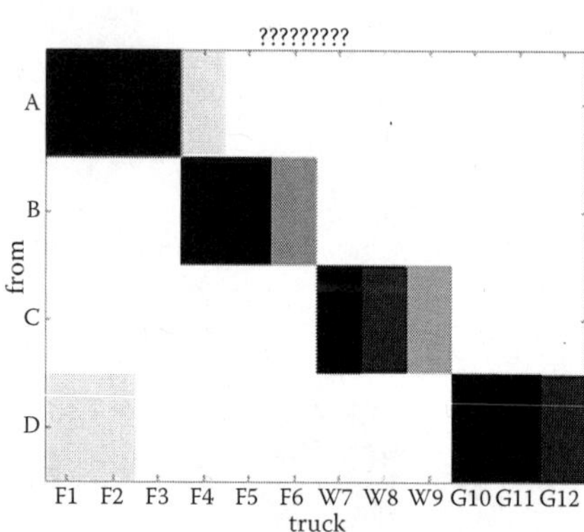
</td></tr>
<tr><td>Partitioned according to "preference"</td><td>The participants command three trucks each. The partition is based on the participants' preferences. This partition requires an active statement from at least one participant in which he/she asks for a specific set of trucks. (In teams with 3 participants, the participants maneuver one truck type each, but not in the order of A: fire trucks 1–6; B: water trucks 7–9; C: gas trucks 10–12.)</td><td>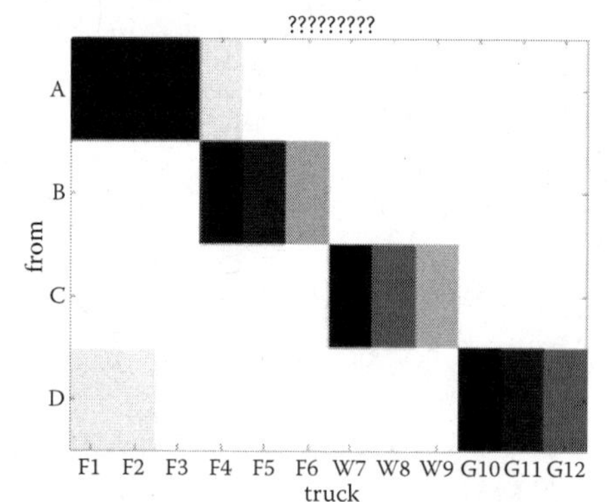
</td></tr>
<tr><td>Assistant</td><td>One participant coordinates the others' actions through e-mail communication and actively commands trucks as he deems appropriate.</td><td>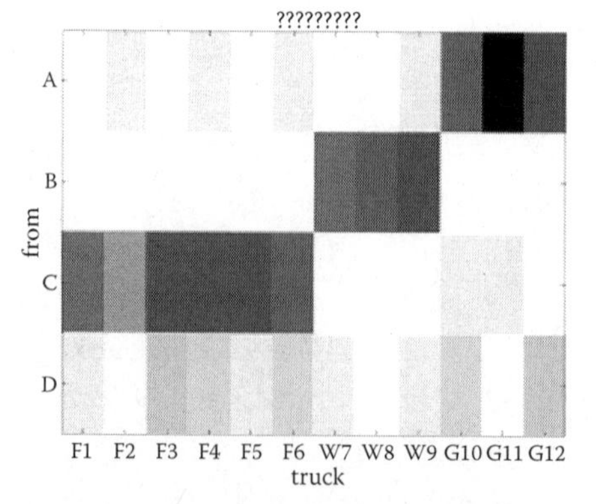
</td></tr>
</table>

Table 3.1 (continued) The Taxonomy of Organizational Structures

CATEGORY	DESCRIPTION	EXAMPLE
Coordinator	One participant coordinates the others' actions through e-mail communication. The leader actively commands trucks occasionally but does not send commands to more than three trucks.	
Shared fire trucks	Two participants command the fire trucks together. The third participant commands the gas trucks, and the fourth commands the water trucks.	
Shared gas trucks	One participant commands all six fire trucks. Another participant commands the water trucks and the other two participants command the gas trucks together.	

?????????
A B C D
from
F1 F2 F3 F4 F5 F6 W7 W8 W9 G10 G11 G12
truck

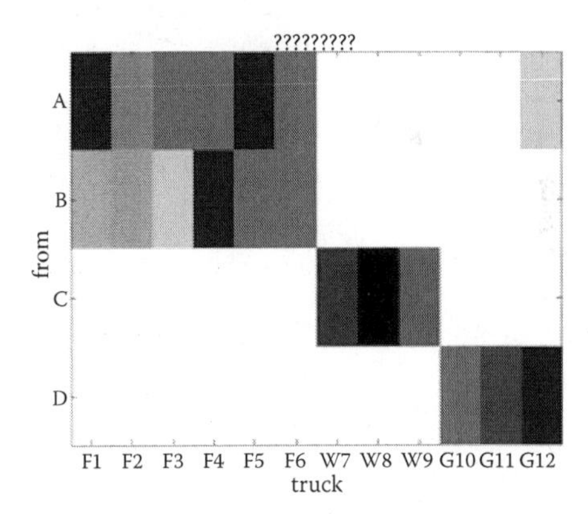

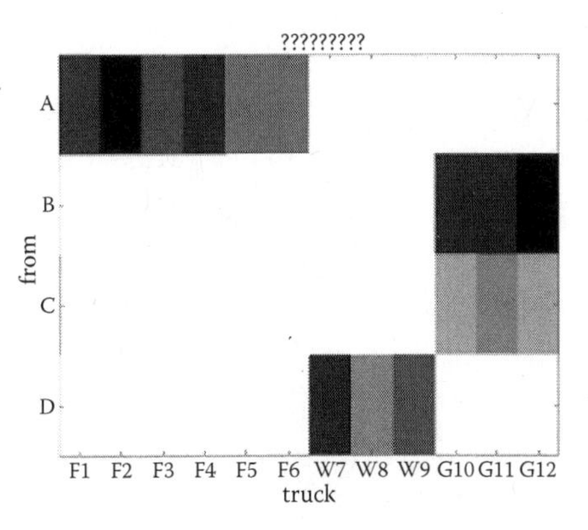

(continued on next page)

Table 3.1 (continued) The Taxonomy of Organizational Structures

CATEGORY	DESCRIPTION	EXAMPLE
Open structure	There is no visible structure. Most participants send commands to a large number of trucks.	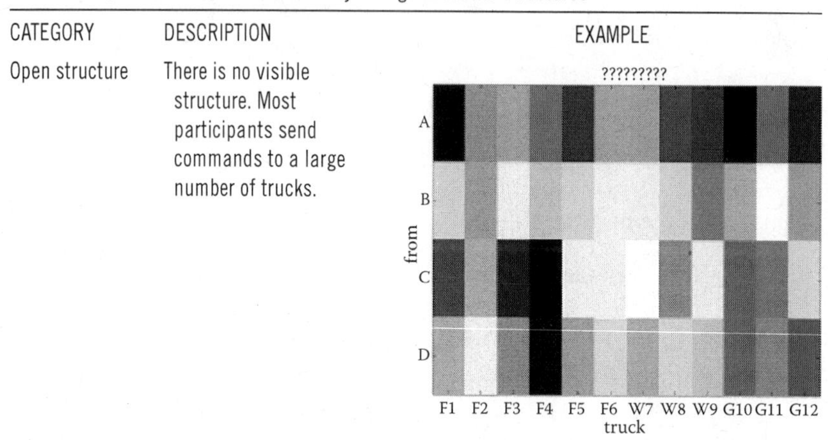

Table 3.2 Counts and Frequencies of Categories of Organizational Structure across the National Groups

CATEGORIES	SWEDES	%	BOSNIANS	%	PAKISTANIS	%	INDIANS	%
Partitioned								
by convenience	37	57.8	5	10.9	14	22.6	2	3.2
by preference	8	12.5	15	32.6	12	19.4	14	22.6
Hierarchic								
Assistant	12	18.8	3	6.5	2	3.2	7	11.3
Coordinator	3	4.7	0	0.0	1	1.6	2	3.2
Shared								
fire trucks	4	6.3	0	0.0	10	16.1	0	0.0
fuel trucks	0	0.0	0	0.0	5	8.1	4	6.5
Open	0	0.0	23	50.0	18	29.0	33	53.2

structures. An analysis comparing the use of partitioned structures to the use of all other structures reveals unexpectedly frequent use by our Swedish teams and unexpectedly rare use by our Indian teams, $\chi^2(3, N = 234) = 26.0, p < .001$.

The shared organizational structure was much more popular with the Pakistani teams than with the other three national groups, $\chi^2(3, N = 234) = 21.2, p < .001$. It appears that our Pakistani teams were uniquely willing to work in a relatively unstructured but fully collaborative manner.

No Swedish matrix was categorized as an open structure. In sharp contrast, more than 50% of Bosnian and Indian teams appear to have preferred the open structure. During these trials, everyone drove a little bit of everything. It is not clear from these data whether the cause was that our groups of Indians and Bosnians distrusted organization, were truly cooperative, or were comfortable with spontaneous chaos. However, it is clear that teams drawn from these two national groups were ready and willing to respond flexibly to the dynamic situation generated by C3Fire.

This analysis suggests that the Bosnians and Indians tended to adopt much more flexible structures than the Swedes, and that the Pakistani teams adopted one of the two shared approaches much more often than the other groups. We use the notation (S // P // BI) to represent a pair of cultural faultlines and three alignments of norms: (a) the alignment of the Open organizational structure preferred by our Bosnian and Indian teams, (b) the disparity between that structure and the structures favored by the Swedish and Pakistani teams, and (c) the uniquely Pakistani tendency to adopt, on occasion, a Shared structure.

3.5.3 Feedback

Our third measure of team collaboration is derived from the e-mails that team members sent to each other using the communication tool built into C3Fire. During the trials this was their only mechanism for cooperation. C3Fire captures a record of all e-mails sent, including the time, sender, and to whom the message was sent. We scored these protocols for two valences of feedback, positive and negative. Positive feedback was defined as a message meant to enhance camaraderie, and negative feedback was defined as a message meant to disparage a player, the team, or its performance. A typical positive message was "Good show!" Negative feedback was often a more colorful and explicit indication of impatience, displeasure, or bad attitude.

To quantify cultural differences in the use of feedback we conducted two analyses. The first concerned the relative frequency of all communication that was classified as feedback, either positive or negative. The two-way ANOVA (group X trial) indicates that national group was significant, $F(3, 202) = 7.82$, $MSE = 0.062$, $p < .001$, power $> .98$. Experimental trial and the interaction of group and trial were not found to be significant. The Tukey HSD test indicates that (a) the Indians and Pakistanis sent significantly fewer feedback messages than both the Swedes and the Bosnians, that (b) the Swedes and Bosnians did not differ from each other, and that (c) the Indians and Pakistanis did not differ from each other (SB // PI). This cultural faultline separates the Europeans from the Asians.

The second analysis on feedback concerned the relative frequency of positive and negative feedback. The Swedes sent few negative statements. In contrast, approximately half of the feedback sent by the other three national groups was negative. A two-way ANOVA (group X trial) indicated that national group was significant, $F(3, 202) = 7.30$, $MSE = 1.148$, $p < .001$, power > .95. Experimental trial and the interaction of group and trial were not found to be significant. The Tukey HSD test indicated that the Swedes differed significantly from the other three groups and that those groups did not differ from each other (S // BPI). This single faultline may reflect a Swedish tendency to be polite at all times or a reticence to engage in necessary confrontation or both. Regardless of interpretation, it is clear that the three other national groups did not practice Swedish reserve.

3.5.4 Caveat

It is important to remember that our aim was to identify prototypical cultural differences in norms for team behavior that may pose barriers to cohesion in multinational teams. We do not pretend to have provided a cultural map of these four specific national groups. The particular differences presented here are less interesting than the fact that the microworld methodology makes them easy to elicit. We present these dimensions of cultural diversity and the resulting faultline models as general exemplars of the varied barriers to effective teamwork that are likely to appear whenever multinational teams are formed.

3.6 Implications for Etiquette in the Design of Technology

Participants in our study worked in culturally homogeneous teams (Swedes with Swedes, etc.). They were matched by age, education, gender, and familiarity with computers and computer-mediated communication tools. All teams were given the same mission and free rein over how to accomplish it. It is therefore compelling to find that

Table 3.3 Dimensions of Cultural Diversity and Faultline Representation

Goal setting	S	//	B		P	//	I
Organizational structure	S	//	P	//	B		I
Positive/negative feedback	S	//	B		P		I
Total feedback	S		B	//	P		I

Note: S = Swedes, B = Bosnians, P = Pakistanis, I = Indians. // = Cultural faultline.

the four different national groups each chose to approach the task so differently.

Table 3.3 summarizes the four patterns of alignment and disparity reported here and the corresponding set of cultural faultlines. The metric for fire truck density identified three different goals and formed a pair of cultural faultlines. The major rift separates the Indians from the other three national groups. A lesser fault separates the Swedes. It remains to be seen whether this diversity in goal setting generalizes beyond the teams we studied and the C3Fire microworld. Nevertheless, it has a serious implication: Multinational teams are vulnerable to the formation of multiple faultlines in norms for goal setting. These faultlines are, in turn, likely to become loci of friction and become barriers to team collaboration. At the extreme, the faultlines could splinter the team into culturally homogeneous subgroups. Diversity in goal setting is a recipe for team dysfunction.

For the designers of technology that will be used by multinational teams, it is not safe to assume that everyone on the team has equivalent expectations about the team's goal. Rather, designers of technology for multinational teams must assume that people from different cultures have diverse norms for goal setting. Further, they must also assume that it is likely that people from seemingly diverse cultures (e.g., Bosnia and Pakistan) will embrace the same norms for goal setting.

Technology that is intended to facilitate multinational teams in goal-directed action must elicit and adjudicate the establishment of and adherence to the team's goals. For well-defined tasks this should not be an onerous duty. Designers of technology that will be employed in novel domains must either conduct or borrow a means-ends task analysis or some other formalized explication of the team's options to assist its efforts to settle upon a goal structure.

Etiquette is part of the remedy. The software that facilitates goal setting and adherence must employ etiquette that emphasizes that there is no one criterion for "best" that can be used to assess a team's selection of goals and its subsequent performance. Every effort should be made to make the elicitation of goals as nonjudgmental and as free from cultural bias as possible. That said, the same functionality must also stress that the goal structure that the team adopts establishes the criteria by which its performance will be evaluated. This push and pull makes it politic to design functionality that elicits what the team (feels it) can do before it attempts to match those capabilities to the options revealed by the task analysis. Technology that employs etiquette does not impose goals; it fits goals to capabilities and the propensities of cultural norms.

Our data on organizational structure support similarly broad inferences. Here, there are two faultlines. One isolates the Swedes, the other, the Pakistanis. The Bosnians and Indians share a common ground. The three norms for organizational structure—strictly partitioned roles, shared roles, and flexible roles—are so different that it is easy to imagine that a multinational team might be spontaneously combustible. The offset along these faultlines is sufficiently great to support the prediction that divergent norms for organizational structure are likely to be a salient factor in the breakdown of many multicultural teams. The road to bridging these faultlines is familiar: Technology that mediates a multinational team must promote cooperation and agreement regarding members' roles and responsibilities. It must also monitor the team to ensure continued compliance throughout the life of the team.

The etiquette associated with promoting agreement on roles and responsibilities is more a function of the several cultures that constitute the team than it is a function of the roles themselves. Eliciting preferences for roles is an instance of elicitation generally. People from different cultures are comfortable with different processes for making their wishes and preferences known (e.g., Duranti, 1997; Hofstede, 1980; Kim and Markus, 1999; Klein, 2005; P. B. Smith et al., 2006). The discussion in Chapter 1 in this volume enumerates several social and cognitive dimensions of cultural diversity that must be respected in any elicitation process.

Our results on the use of feedback in e-mail communication indicate that acting naturally is rather likely to offend. It is clear that the other national groups did not possess Swedish reserve. The Swedish aversion to negative comments might lead them to think poorly of team members who expect off-hand criticism to be taken lightly. Conversely, Swedes might lose respect by failing to be critical when others expect it is due. This and other culturally-driven differences in communication style could readily be misinterpreted in a newly-formed multinational team.

Technology that seeks to diffuse the offense that one team member might take in response to another's spontaneous behavior cannot censor or censure that behavior. Rather, it must illuminate the cultural context in which that behavior is seen as normal and appropriate. This contextualization of behavior cannot be undertaken in the midst of time-critical operations. There is simply no time to salve wounded feelings in the heat of action. Contextualization must be part of an education process during team formation.

3.7 Summary

Different national cultures have different norms for teamwork. These differences are profound and have severe implications for the design of technology and the application of etiquette in that technology. The observed multiplicity of cultural faultlines is likely to be the rule rather than the exception. Whenever people from different cultures are thrown together to form a team, it is likely that there will be multiple patterns of alignment and disparity. The other team members may be like you in some ways and unlike you in others. Accordingly, the designers of technology to support multinational teams should be prepared for a swarm of cultural faultlines.

Designers of technology for multinational teams need to recognize that etiquette that seeks to bridge cultural faultlines plays a role at all stages of a team's existence. Technology that seeks to illuminate alternative norms for feedback must be used during training. Once the team is actually assembled, it needs to establish its goals. We suggest that it might be politic to match performance capabilities to a formal task analysis during the goal formation process. Finally, the methods used to elicit preferences must be sensitive to dimensions of social and cognitive diversity that have been well-documented by researchers in the field of cross-cultural psychology.

Acknowledgments

Much of the data were collected at Skövde University. The hospitality and assistance of our colleagues in Skövde are greatly appreciated. This research was supported by a grant from the Swedish Rescue Services Agency. Per Becker and Bodil Karlsson were truly helpful project monitors. Igor Jovic and Syed Zeeshan Faheem assisted in the administration of the experiments and the translation and coding of the Bosnian and Pakistani e-mail communication, respectively. The authors thank Magnus Bergman, Helena Granlund, Paul Hemeren, Lauren Murphy, Lars Niklasson, Erik Ohlsson, Milan Veljkovic, and Rogier Woltjer for their invaluable assistance during various phases of the project.

References

Brehmer, B. (2005). Micro-worlds and the circular relation between people and their environment. *Theoretical Issues in Ergonomics Science*, 6 (1), 73–93.

Brehmer, B. and Dörner, D. (1993). Experiments with computer-simulated microworlds: Escaping both the narrow straits of the laboratory and the deep blue sea of the field study. *Computers in Human Behavior*, 9 (2–3), 171.

Duranti, A. (1997). *Linguistic Anthropology*. Cambridge, England: Cambridge University Press.

Granlund, R. (2002). Monitoring distributed teamwork training. Unpublished doctoral dissertation, Linköpings Universitet, Linköping, Sweden.

Granlund, R. (2003). Monitoring experiences from command and control research with the C3Fire microworld. *Cognition, Technology and Work*, 5 (3), 183–190.

Gray, W. D. (2002). Simulated task environments: The role of high-fidelity simulations, scaled worlds, synthetic environments, and laboratory tasks in basic and applied cognitive research. *Cognitive Science Quarterly*, 2, 205–227.

Hofstede, G. (1980). *Culture's Consequences: International Differences in Work-Related Values*. London: Sage Publications.

Hofstede, G. and Hofstede, G. J. (2005). *Cultures and Organizations: Software of the Mind.* (2nd ed). London: McGraw-Hill.

Johansson, B., Persson, M., Granlund, R., and Mattsson, P. (2003). C3Fire in command and control research. *Cognition, Technology and Work*, 5 (3), 191–196.

Kim, H. and Markus, H. R. (1999). Deviance or uniqueness, harmony or conformity? A cultural analysis. *Journal of Personality and Social Psychology*, 77 (4), 785–800.

Klein, H. A. (2004). Cognition in natural settings: The cultural lens model. In M. Kaplan (Ed.), *Cultural Ergonomics: Advances in Human Performance and Cognitive Engineering* (pp. 249–280). Oxford: Elsevier.

Klein, H.A. (2005). Cultural differences in cognition: Barriers in multinational collaborations. In H. Montgomery, R. Lipshitz, and B. Brehmer (Eds.), *How Professionals Make Decisions* (pp. 243–253). Mahwah, NJ: Lawrence Erlbaum Associates.

Lau, D. C. and Murnighan, K. J. (1998). Demographic diversity and faultlines: The compositional dynamics of organizational groups. *Academy of Management Review*, 23 (2), 325–340.

Lau, D. C. and Murnighan, K. J. (2005). Interactions within groups and subgroups: The effects of demographic faultlines. *Academy of Management Review*, 48 (4), 645–659.

Lindgren, I. and Smith, K. (2006). Using microworlds to understand cultural influences on distributed collaborative decision making in C2 settings. *Proceedings of the 11th Annual International Command and Control Research and Technology Symposium (ICCRTS)* (CD-Rom). Cambridge, UK.

Matsumoto, D. (2003). The discrepancy between consensus-level culture and individual-level culture. *Culture and Psychology*, 9 (1), 89–95.

Molleman, E. (2005). Diversity in demographic characteristics, abilities and personal traits: Do faultlines affect team functioning? *Group Decision and Negotiation*, 14, 173–193.

Rigas, G., Carling, E., and Brehmer, B. (2002). Reliability and validity of performance measures in microworlds. *Intelligence*, 30, 463–480.

Schwartz, S. H. (1992). Universals in the content and structure of values: Theoretical advances and empirical tests in 20 countries. *Advances in Experimental Social Psychology*, 25, 1–65.

Smith, K., Lindgren, I., and Granlund, R. (2008). *Exploring Cultural Differences in Team Collaboration*. Manuscript submitted for publication.

Smith, P. B. and Bond, M. H. (1999). *Social Psychology across Cultures*. London: Prentice Hall Europe.

Smith, P. B., Bond, M. H., and Kağitçibaşi, Ç. (2006). *Understanding Social Psychology across Cultures: Living and Working in a Changing World*. London: Sage Publications.

Strohschneider, S. and Güss, D. (1999). The fate of the Moros: A cross-cultural exploration of strategies in complex and dynamic decision making. *International Journal of Psychology*, 34 (4), 235–252.

Thatcher, S. M. B., Jehn, K. A., and Zanutto, E. (2003). Cracks in diversity research: The effects of diversity faultlines on conflict and performance. *Group Decision and Negotiation*, 12, 217–241.

Triandis, H. C. (1996). The psychological measurement of cultural syndromes. *American Psychologist*, 51 (4), 407–415.

PART II

Introducing Etiquette and Culture into Software

4

Computational Models of Etiquette and Culture

PEGGY WU, CHRISTOPHER A. MILLER, HARRY FUNK, AND VANESSA VIKILI

Contents

> Do not flatter or wheedle anyone with fair words, for he who aspires to gain another's favor by his honied words shows that the speaker does not regard him in high esteem, and that the speaker deems him far from sensible or clever, in taking him for a man who may be tricked in this manner.
>
> **—English translation from Maxims II (27), reprinted in George Washington's Rules of Civility***

4.1 Introduction

Etiquette is often defined as a shared code of conduct. Social etiquette such as how to greet your new boss from Japan can be seen as a discrete set of rules that define the proper behaviors for specific situational contexts. Those who share the same etiquette model (i.e., the same rules and interpretations of these rules) may also share expectations of appropriate behaviors, and interpretations of unexpected behaviors. When people lack a shared model of etiquette, the result may be confusing, unproductive, or even dangerous interactions.

Etiquette is a well studied phenomenon in linguistics and sociology; it is a vital part of communication in virtually all cultures and all types of interactions. However, what is of particular interest in the work reported throughout this volume, is the increasing evidence that people expect their interactions with computers to be very like their interactions with humans. Not only do people readily anthropomorphize technological artifacts, but they relate to them at a social level (Reeves and Nass, 1996). We propose that the concepts of etiquette can be expanded and used to design effective human–computer interactions, and predict human reactions to computer behaviors. In this chapter we present a well studied and influential body of work on politeness in human interactions, discuss its validity in human–computer interactions, present a "believability" metric, and provide examples of empirical experiments quantifying the implementations to show that models of etiquette are not only amendable to quantitative modeling and analysis, but can also predict human behavior

* Washington, George (2003) George Washington's Rules of Civility, fourth revised Collector's Classic edition, Goose Creek Productions, Virginia Beach, VA.

in human–computer and computer mediated interactions. We briefly discuss a computational model of etiquette and a series of experiments that empirically validate this model, and conclude by discussing challenges of current models and future work.

4.2 A Model of Politeness and Etiquette

Brown and Levinson (1987) produced a seminal body of sociological and linguistic work on politeness. In this work, they developed a model of politeness from cross-cultural studies. They noted that regardless of language or culture, people regularly deviated from what is considered "efficient speech," as characterized by Grice's (1975) *conversational maxims.* Grice's rules of efficient speech consist of the "Maxims of Quality" (i.e., contain truthfulness and sincerity), quantity (i.e., are concise), relevance (i.e., have significance to the topic at hand), and manner (i.e., have clarity and avoid obscurity). For example, the word "please" is appended to a request such as "Please pass the salt." The use of "please" is unnecessary for a truthful, relevant, or clear message, and it violates the Maxim of Quantity since it adds verbiage.

Brown and Levinson collected and catalogued a huge database of such violations of efficient conversation over a period of many years in cross-linguistic and cross-cultural studies. Their explanation for these violations is that additional "polite" verbiage may be necessary to clarify ambiguities inherent in human communications, such as the relationship between the speaker and hearer, and context of the communication.

The Brown and Levinson model assumes that social actors are motivated by two important social needs based on the concept of *face*, which has both positive and negative facets (Goffman, 1967). *Positive face* is associated with an individual's desire to be held in high esteem and to be approved of by others, while *negative face* is related to an individual's desire for autonomy. Virtually all interactions between social agents involve some degree of threat to each participant's face. Brown and Levinson call actions that produce face threats *face threatening acts* (FTAs). Even when there is no power difference between two actors, a speaker inherently places a demand on a hearer's attention by the simple act of speaking, thus threatening the hearer's negative face. If the speaker simply states a request such as, "Give me the salt," the intent may have been communicated efficiently, but the speaker has been ambiguous about whether or not the speaker has the right to compel the hearer to comply. By adding the word "please" the speaker attempts to communicate acknowledgment of his or her lack of power to demand the hearer's compliance, thereby *redressing* or mitigating the threat implicit in the request.

4.2.1 Computing the Severity of a Face Threat

The core of Brown and Levinson's model is the claim that the degree of face threat posed by an act is described by the function:

$$W_x = D(S,H) + P(H,S) + R_x \tag{4.1}$$

where:

- W_x is the "weightiness" or severity of the FTA; the degree of threat.
- $D(S,H)$ is the *social distance* between the speaker (S) and the hearer (H). It decreases with contact and interaction, but may also be based on factors such as membership in the same family, clan, or organization.
- $P(H,S)$ is the *relative power* that H has over S.
- R_x is the *ranked imposition* of the act requested, which may be culturally influenced.

To avoid disruptions caused by FTAs, people use *redressive strategies* to mitigate face threat imposed by their actions. Brown and Levinson claim that if the social status quo is to be maintained between speaker and hearer, then the combined value of the redressive strategies used should be in proportion to the degree of the FTA. That is:

$$W_x \approx V(A_x) \tag{4.2}$$

where $V(A_x)$ is the combined redressive value of the set of the redressive strategies (A_x) used in the interaction.

Brown and Levinson offer an extensive catalogue of universal strategies for redress, organized and ranked according to five broad strategies described below from least threatening to most.

1. **Refraining from carrying out the FTA**. The least threatening approach is simply to not carry out the FTA, which is sometimes the only acceptable strategy in some cultural contexts. For example if one is cold, to simply put up with being cold, rather than confronting the hearer with a request to turn up the heat.
2. **Off record**. If the FTA is carried out, then the least threatening way to do it is "off record," that is, by means of innuendo and hints, without directly making a request, e.g., "It is a bit cold in here."
3. **Negative redressive strategies** focus on the hearer's desire for autonomy and attempt to minimize the imposition on the hearer. Examples of these strategies include being direct and simple in making the request, offering apologies and deference (e.g., "If its not too much trouble, would you mind turning up the heat?"), minimizing the magnitude of the imposition (e.g., "Could you turn up the heat just a little bit?"), or explicitly incurring a debt (e.g., "Could you do me a favor and turn up the heat?")
4. **Positive redressive strategies** acknowledge the hearer's desire for approval by emphasizing common ground between speaker and hearer, e.g., "We are freezing in here. Let's turn up the heat." Other examples include the use of honorifics, in-group identity markers, nicknames or jokes, and the display of sympathy.
5. **Bald.** Finally, the most threatening way to performing an FTA is "baldly" and without any form of redress, e.g., "Turn up the heat."

4.2.2 Validity for Use as Computational Model

Brown and Levinson's model provides a taxonomy of linguistic strategies that collectively enable a speaker to convey information, mitigate face threats, and predict the degree of face threat that must be mitigated based on the context. Their model can provide a framework for more detailed computational models because it has a solid grounding in empirical research that few other models can claim. The components of the model are also largely orthogonal. While there can be interactions between the components of Brown and Levinson's model, the components can be evaluated separately. For example, power difference may have complex implications with social distance; consider the situation in which one is friends with a superior. However, power and social distance can be assessed separately, and their interaction represented outside of this model.

Brown and Levinson's model has been criticized as overly simplified, Anglo-centric, and not universally applicable (Eelen, 2001; House, 2005). For example, it does well characterize sarcasm and irony. However, its simplicity is also its strength when used in a computational model from a practical engineering standpoint. Calculating face threat requires relatively few computing resources, even in simulations involving many individual agents, such as crowd control scenarios.

In our work, we have focused on a specific type of speech act called *directives*, which are requests to perform a task (Searle, 1969). We feel that Brown and Levinson's model adequately assesses the language most often used in directives, and forms a strong foundation upon which constructs can be added if needed. Studies of human interactions have shown that this model can be adapted for quantitative modeling and analysis (Shirado and Isahara, 2001), suggesting that it can also be adapted for use as a computational model to guide appropriate computer behavior in human–computer interactions.

4.3 Application of Etiquette to Human–Machine Interactions

Anecdotal and empirical evidence support the theory that humans are not only capable of social interactions with machines, but that they do so naturally. Nass (Reeves and Nass, 1996; Nass, 1996) conducted a series of experiments demonstrating that humans readily generalize patterns of social conduct to human–computer interactions. He calls this relationship "the media equation." We claim that models of etiquette not only provide insights into human social interactions, but they can also be used to inform human–machine interactions and predict human perceptions of those interactions.

The notion of *selective-fidelity* for simulations (Schricker et al., 2001) places focus on the aspects of a situation that make a functional difference from the user's perspective. We believe giving computer agents the ability to exhibit politeness and etiquette will provide the biggest "bang for the buck" for increasing the realism of computer agents, even more so than giving them more convincing and human-like physical appearances. In robotics, Mori (1970) formulated a concept called the *uncanny valley* to describe human reactions to the physical appearance of robots. He hypothesized that as robots become increasingly human in appearance, our emotional response towards them becomes increasingly positive. However, when a robot's appearance becomes too close to human without exactly replicating it, our emotional response abruptly turns negative and we feel revulsion, much as we would towards corpses or zombies. The uncanny valley describes the dramatic dip in an emotional response graph that occurs as the robot's appearance approaches human. We hypothesize that a lack of *believable* social interactions may produce a similar plunge into an uncanny valley that is independent of the agent's physical realism. After all, we generally do not feel discomfort interacting with people with physical deformities, but we do feel discomfort when actors are inappropriately rude, overly polite, or do not react in the expected ways to rudeness or politeness. Interactions that consistently violate etiquette norms may cause unease and cognitive dissonance, even if the actor's appearance is perfectly human.

One of our motivations for developing a computational model of etiquette is to make it possible to evaluate whether an agent's interactions are believable. We speculate that in general, utterances that have large deviations from their expected range of redress (e.g., perceived as extremely rude or extremely polite) will have detrimental effects on performance. We claim that when a behavior becomes so extreme as to provoke an "unbelievable" response from the hearer, the resulting cognitive dissonance will be severe enough to increase his or her workload and interrupt ongoing tasks, thus harming performance metrics. To avoid this undesirable effect, a computer agent must exhibit believable behaviors. How, then do we evaluate and formulate believable behaviors? Bates (1994) describes the use of traditional animation techniques to build seemingly emotional and therefore believable agents. He summarizes Thomas and Johnston's (1995) key points for creating "the illusion of life": clearly define emotional states, show emotions revealed in the thought process of agents, and accentuate emotion through exaggeration and other storytelling techniques. We believe etiquette can facilitate the portrayal of emotions by serving as a bridge between thought process and behaviors exhibited by the agent. Human etiquette rules and theory are multifacetted and subjective. As such, the challenge of evaluating believability in a computational manner is to develop a model that utilizes concrete variables to embody the vast array of beliefs and situational contexts. In the following sections, we will examine work that models social attributes computationally, describe the authors' efforts to create a computation model formalizing the concept of believability, and provide examples of its application.

4.4 Examples of Computational Models of Etiquette

Some of the earliest computational applications of politeness in computer science literature can be found in dialog systems. Pautler (1998) developed a taxonomy of perlocutions, the impact of speech acts on social attitudes and behaviors, and applied it in a message composing system called LetterGen. LetterGen takes general communication and social goals from the user, such as "decline an invitation politely," and selects socially appropriate speech acts based on those goals. This work combined traditional models from Natural Language Processing (NLP) with social psychology by considering the effect of language on interpersonal relationships. While this work does not provide a formalized mathematical model of politeness, it organizes speech into varying levels of politeness and represents their effects as plan operators. Ardissono et al. (1999) describe a rational model used to represent knowledge, including the politeness with a focus on indirect speech acts (e.g., off-record strategies). This formalism provides the mechanism for reasoning about speech acts and goals in a computer speech recognition system.

Cassell and Bickmore (2002) successfully used a model of social interactions derived from Brown and Levinson's work to actively tune an avatar's behaviors in the domain of real estate sales. The avatar's goal was to build up the human client's trust in the avatar to a level that would allow the avatar to introduce face-threatening topics pertinent to real estate sales—such as family size, marital status, and income level. It does this through the use of *small talk*. *Small talk* is ongoing conversation about nonthreatening topics designed to build familiarity, allowing one to build up to face-threatening topics gradually. Their system uses a computational variation of Brown and Levinson's model to assist in determining the degree of threat posed by specific topics in specific contexts. They have not, however, explicitly applied this approach to evaluating believability or managing performance.

4.5 Brown and Levinson's Model in Human–Computer Interactions

Brown and Levinson's model describes politeness in human social interactions, but do their theories transfer to human–computer interactions? Further, does it matter if people perceive computers to be polite or rude? That is, does politeness displayed by a computer impact human work performance, and if so, what are the implications on how we should design computer assistants? We explored these questions in a series of experiments, the first of which is described in this section.

The authors have been exploring the concept of politeness for use in automation since 2000 (Miller and Funk, 2001; Miller, 2003; Miller et al., 2004; 2005; 2006; 2007). One of our initial efforts examining the role of etiquette in automation has been in the domain of eldercare (Miller et al., 2004). In work funded in part by Honeywell International and the National Institute of Science and Technology, we conducted an experiment to validate Brown and Levinson's model. The application was a smart home system with strategically placed sensors and Web and phone-based user interfaces, designed to enable the elderly to stay in their homes and lead independent lives longer. Medications were stored in a "smart" medication caddy equipped with a sensor that communicated wirelessly with the smart home system. If the caddy has not been opened within a specified time window, it indicated that they had missed their medication, and the system would issue a reminder message over the telephone. Our hypothesis was that polite messages would result in better affect and compliance. However, we first needed to assess whether people would actually perceive a message from a computer to be polite or impolite.

Using Brown and Levinson's general redressive strategies as a guideline, we constructed statements using different types of redressive strategies ranging from off-record, to negative politeness, to positive politeness, to bald. The messages were as follows:

Bald:	You've missed a dose of medication. Take your medication now.
Positive Politeness:	Your health is important. It looks like you've missed a dose of medication you wanted me to check on. Why don't you take your medication now?
Negative Politeness:	I'm sorry, but Med-Advisor hasn't detected you taking your medication schedule for <time>. If you haven't taken it, could you please take it now?
Off-Record:	This is Med-Advisor calling to remind you that your health is important.

The utterances were variations of a message issued to patients by the medication monitoring system when the patient missed a dose of his or her medication. Patients were asked to rate the politeness of each utterance. The goal of this experiment was to determine whether the subjects' perceptions of politeness matched Brown and Levinson's predictions.

Subjects included elders and adults from a wide range of age groups. Some were familiar with the medication reminder system, and some were not. The subjects were presented with all utterances simultaneously either on paper or electronic format, without further information about whether the messages were responses to missed medication events, or they were false alarms. We asked the subjects to rank the utterances in terms of politeness and appropriateness, with no further definition of these terms.

We found that the subjects' rankings were consistent with the Brown and Levinson predictions of politeness for all utterances except the "off-record" case. We feel this may be because off-record strategies can require much subtlety and attention to context to be properly applied and understood. However, like e-mail, a computer-generated text message conveys little context or subtlety, making it a challenging media to convey off-record utterances as intended. This is consistent with House's (2005) observation that the relationship between indirectness and politeness is complicated and context dependent. We conclude from this experiment that it may be best to avoid use of off-record strategies in computer text messages.

4.6 The Impact of Etiquette on Performance in Human–Computer Interactions

If people do perceive statements from computers to be polite or rude in much the same way as they view similar statements from people, does this matter? Does it affect their performance in terms of reaction time, compliance, work load, or other important dimensions? We assessed the impact of etiquette on human performance in a study sponsored by the U.S. Air Force Research Laboratory.

We hypothesized that utterances from computer agents that deviate greatly from the expected level of redress will be perceived as either rude or overly polite (and thus suspect), and will have detrimental effects on performance. We hypothesized that politeness will tend to increase compliance, whereas rudeness will decrease compliance, all other factors being the same. We also hypothesized that there will be a relationship between compliance, trust, and positive effect that comes with expected, pleasing, and/or adequately polite interactions. These hypotheses are based on the concepts that appropriate levels of trust of automation benefits performance (Lee and See, 2004). Parasuraman and Miller (2004) provide some specific experimental data on trust and affect, Norman (2004) on pleasure and affect, and Cialdini (1993) on the relationship between flattery and affect.

Table 4.1 summarizes our hypotheses for the human performance metrics that we believe are influenced by culture, and represents the believable region for redressive behaviors. Note that in all cases we are referring to the etiquette as perceived and expected by an

Table 4.1 Hypothesized Relationships between Etiquette and Performance Dimensions

PERFORMANCE METRIC	DECREASING POLITENESS	NOMINAL	INCREASING POLITENESS
Cognitive workload	Increasing	Nominal	Increasing
Situation awareness for etiquette variables	Increasing	Nominal	Increasing
Compliance	Decreasing	Nominal	Increasing
Trust	Decreasing	Nominal	Increasing
Affect	Decreasing	Nominal	Increasing
Reaction time	Decreasing[a]	Nominal	Increasing

[a] While a slight decrease in politeness may imply urgency and therefore decrease reaction time, a larger decrease in politeness may require more cognitive processing on the part of the hearer, causing reaction time to increase.

observer whose cultural background will inform these interpretations and expectations.

4.6.1 Experimental Methods

Our experiment included a demographic questionnaire, a testbed that asked subjects to respond to simulated requests with varying levels of politeness from various types of people, and a survey to capture their perceptions of politeness of the simulated requests. All subjects completed a general demographics questionnaire at the start of the experiment consisting of the Values Survey Module (VSM94) (Hofstede, 2001), the Culture Dimension Survey (CDS) (Dorfman and Howell, 1988), and our own questionnaire, which was designed to assess the subjects' general perceptions of cultural factors as defined by Hofstede (2001) including power, individualism, gender, and uncertainty avoidance.

Testbed and Experimental Scenario. To create the testbed, we modified the Tactical Tomahawk Interface for Monitoring and Retargeting (TTIMR). We selected it for its realism and flexibility, which allowed us to create diverse scenarios (details of TTIMR can be found in Cummings, 2003). We called our modified testbed the Park Asset Monitoring and Management Interface (PAMMI). We developed a scenario in which a subject would play the role of a dispatcher in a national park. Information requests periodically arrived on his/her screen from field agents who are park staff who are located in different areas of the park. The information requests were made by virtual characters known as Directive Givers (DGs), but subjects were not told whether the DGs were controlled by humans or by the computer. Figure 4.1 shows the request screen where DGs make information requests from the human subject.

The types of information requests include location of park vehicles and destinations. All requests were limited to short (one or two word) text answers. Upon receiving a request, subjects were required to examine the PAMMI status screen, obtain the answer, and type it in a text box to respond to the DG. March 26-29 4.2 shows a screenshot of the PAMMI status screen.

There were four conditions in the experiment, and in each condition, subjects were shown the same PAMMI status screen (Figure 4.2) for information gathering, and the information request screen

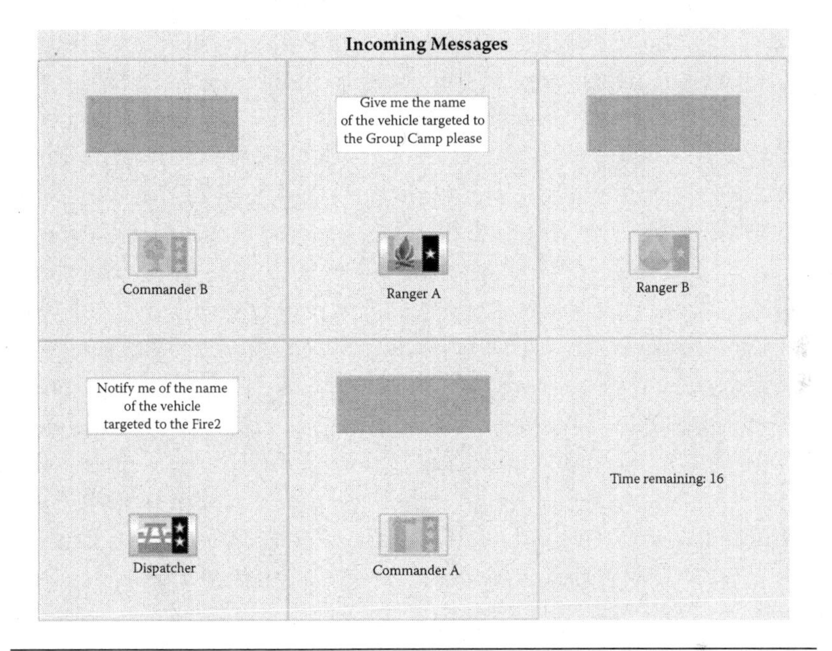

Figure 4.1 The request screen for the power × politeness condition.

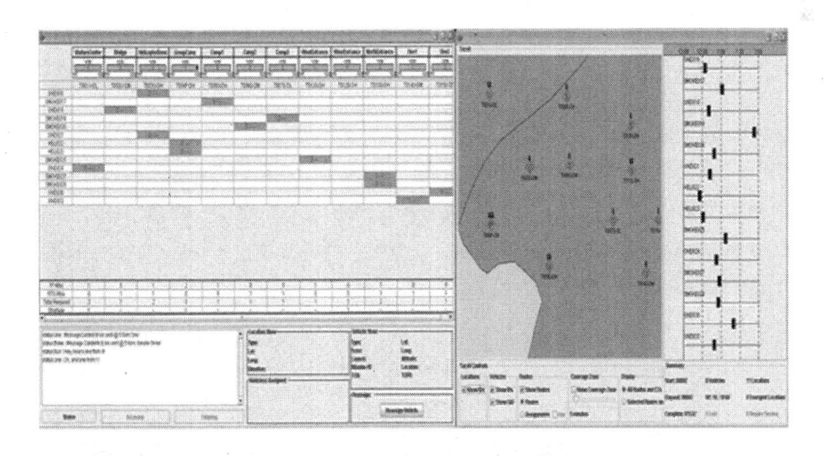

Figure 4.2 The PAMMI status screen showing status of vehicles.

(Figure 4.1) when DGs made a request. In each scenario, there was a total of five DGs who took on different roles, depending on the experimental conditions. The lower right box on the request screen shows the subject how much time he or she has remaining to complete the task. The DGs are represented as simple icons and text messages to reduce subject bias due to factors such as the DGs' tone of voice, accent, and gender (except when we explicitly wanted to show gender). In Figure 4.1, which depicts the "power × politeness" condition, the subject was told that his or her rank is two stars, and that each DG's rank is indicated by the number of stars. The three-star commanders, therefore, have higher power than the subject, while the one-star rangers are lower power and the two-star dispatcher has the same power as the subject. However, the subjects were not told that each DG has a characteristic politeness: always polite, always rude, or neutral. One commander is always polite and one always rude. Similarly, one ranger is always polite, while the other is rude. The dispatcher is always neutral. In Figure 4.1 Ranger A is making a polite request because he or she has added "please" to the end of the request. The dispatcher's request is neutral, neither polite nor impolite.

Subjects. There were 62 subjects aged 18–65 who were recruited from local universities, and the general population completed the study.

Conditions. Subjects were randomly assigned to one of four groups.

- Power × politeness
- Familiarity × politeness
- Gender × politeness
- General politeness

We modified the stimuli for each group by embedding power, group identity, or gender markers in the training materials that the subjects received, and in the icons used for the DGs in the testbed. Figure 4.3 shows icons for the familiarity × politeness and gender × politeness conditions.

Procedure. Subjects were first asked to complete a training session, and then complete a 45-minute session with the testbed where requests

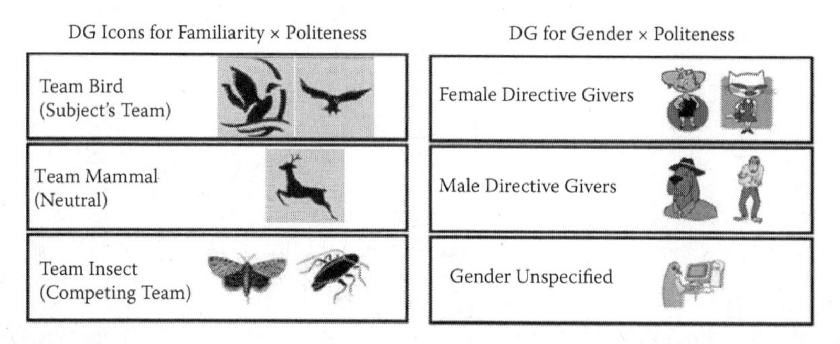

Figure 4.3 Icons for the familiarity × politeness and gender × politeness conditions.

arrived at the rate of one per minute. Of these requests, 25 were "single requests" from one DG at a time (5 from each DG), and the remaining 20 requests were "double requests" in which two different DGs made simultaneous requests. Double requests always came from the neutral dispatcher and one other DG. Figure 4.1 shows an example of a double request from a dispatcher and Ranger A.

For the single requests, subjects were asked to read the question, locate the answer, and reply by entering a response in a text box. For double requests, subjects were asked to read the two questions from two DGs, select the DG to whom the subject wishes to respond, locate the answer, and enter a response in a text box.

During each session, we recorded objective performance metrics including compliance, accuracy, and reaction times. After the completion of each scenario we had subjects complete questionnaires employing self-reported metrics of perceived politeness of DG, effect, trust, perceived competence of DG, and perceived workload caused by DG.

4.6.2 Results

4.6.2.1 Familiarity × Politeness Factorial within-subject ANOVA analyses were carried out for the study of familiarity (the inverse of social distance) × politeness on performance ($n = 20$). Figure 4.4 (left) shows that subjects were significantly more compliant with requests from DGs with higher familiarity than lower familiarity ($p < 0.01$). Further, subjects were significantly more compliant with polite DGs ($p < 0.01$). No significance effect was found for total reaction time (time for subjects to read and acknowledge the request, find the answer, and respond to the DG) (see Figure 4.4 [right]). However, a partial eta squared analysis indicated that a larger sample size may in fact yield significant results. Figure 4.4 (right) shows that subjects may have been slightly more accurate when responding to rude DGs, although no statistically significant effects were detected.

4.6.2.2 Power × Politeness Similar analyses were carried out for the objective metrics for power × politeness ($n = 19$). As predicted, the relative power of the DG was found to have a significant effect on compliance ($p < .01$) (see Figure 4.5[left]). High powered DGs elicited an

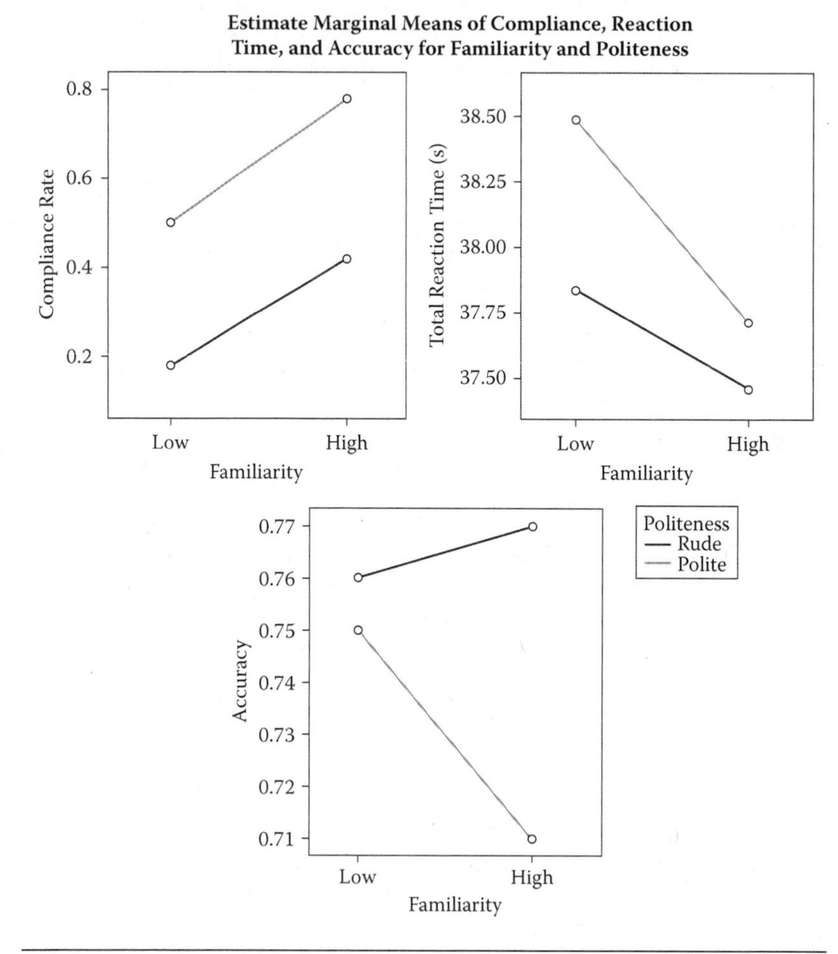

Figure 4.4 Estimated marginal means of compliance (left), reaction time (right), and accuracy (center) for familiarity and politeness.

average compliance rate of 81%, whereas that of low powered DGs was 26%. Polite DGs received slightly higher compliance rates, but not of statistical significance. As Figure 4.5 (right) illustrates, there were no significant effects on total reaction time. However, responses to rude DGs had higher accuracy rates ($p = 0.29$) though, again, this did not reach statistical significance (see Figure 4.5 [center]).

Figure 4.6 (left) shows that when there were two simultaneous requests, one from a high powered DG and one of equal power to the subject, subjects took a significantly more time to decide which DG to select, regardless of the politeness of the request ($p < 0.05$).

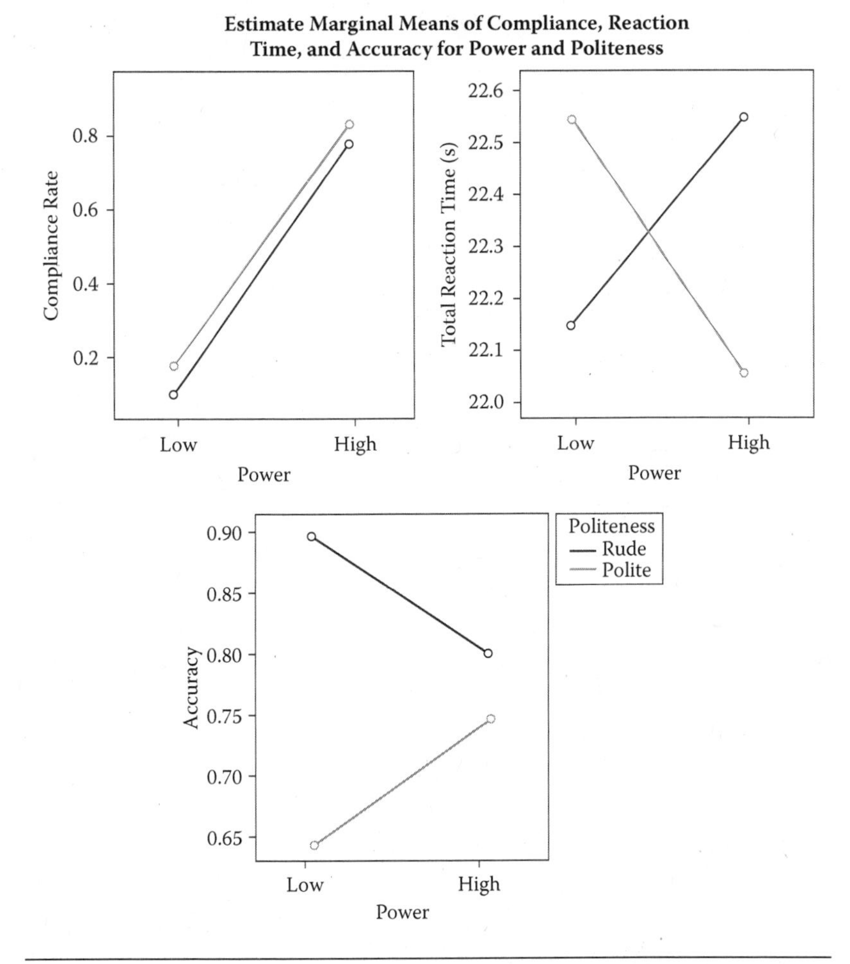

Figure 4.5 Estimated marginal means of compliance (left), reaction time (right), and accuracy (center) for power and politeness.

Figure 4.6 (right) shows that although politeness effects were not significant ($p = 0.29$), subjects took longer to respond to rude DGs even after they found the answer to their request.

4.6.2.3 Gender × Politeness There were several significant effects of gender on task performance. Male DGs produced higher compliance rates as well as shorter reaction times. Further, for subjective measures, male DGs were perceived as more polite than females, even though in the experimental stimuli, the rude male DGs used the same utterances as rude females, and the polite DGs used the same utterances as

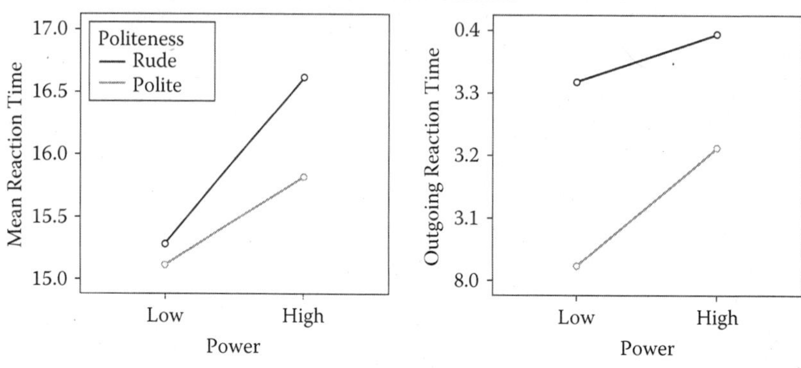

Figure 4.6 Estimated marginal means of detailed reaction times for power and politeness: (left) reaction time for selecting a DG when two simultaneous requests occurred and (right) response time when a single request occurred.

polite females. One speculation as to why females were perceived as more rude is that as a society, we expect females to be more polite, thus when they use an objectively equal amount of politeness as their male counterparts, their politeness level falls short of expectations and they are perceived as more rude.

4.6.2.4 General Politeness Subjective ratings were collected immediately after the testbed session to assess how subjects perceived the politeness of DGs, as well as their affect and trust towards DGs, the perceived workload caused by DGs, and the competence of DGs. Subjects (n = 21) were asked to rate each DG for these variables on a scale from –5 to 5. Perceived politeness was significantly lower for rude DGs ($p < 0.001$). This provides validation that subjects perceived the politeness used in the directives as intended in the design of the DGs. Ratings of both affect and trust were also significantly lower for rude DGs ($p < 0.001$ for both variables), regardless of their power. Politeness was also found to affect perceived workload ($p = 0.058$). Rude DGs made subjects feel as though they were working harder, regardless of power differences. Perceived competence was not found to be significantly affected by politeness. However, marginal means show that competence was perceived to be lowest for rude, low power DGs.

4.6.3 Discussion

These findings generally are consistent with anecdotal evidence and intuition about how people tend to react to familiarity, power, and politeness. Subjects were more compliant with familiar, high-powered, or polite individuals. Interestingly, there was a significant politeness effect in the time in took subjects to select a DG when simultaneous requests occurred. Subjects took more time to decide which DG to select when there was a high-powered neutral DG pair as opposed to a low-powered neutral DG pair. We speculate that subjects may have attributed more importance to the messages from high-powered DGs and were more conscientious in reading their requests, resulting in higher reaction times. As expected, subjects reported that they liked and trusted polite DGs more. Further, subjects felt that rude DGs increased their workload regardless of power or familiarity. It is surprising that we did not find effects of power on perceived workload despite its measured effects on reaction time. This may indicate that the subjects were less likely to view a time-consuming request as an imposition when coming from a high-power DG.

These findings are not novel to cross-cultural studies or general sociology, but instead, demonstrate the feasibility of collecting objective metrics in disciplines that are highly dependent on subjective data and self-reports. We have shown that humans respond to etiquette language even in low-fidelity simulations such as text-based chat. Further, we have provided evidence that such responses can be measured in quantifiable ways in terms of task performance.

4.7 Extending Brown and Levinson: A Believability Metric

Two people can listen to the same interaction and perceive completely different levels of politeness, depending on their individual beliefs, culture, values, and understanding of the situation. The Brown and Levinson model does not directly account for individual differences in perceived politeness. However, it is critical for soldiers, diplomats, business people, or tourists interacting with people from other cultures to understand how such differences can result in accidental rudeness, misunderstandings, and breakdown of communications. For business people, such breakdowns can be financially disastrous, but for soldiers they can be life threatening. In order to model these differences for use in a cross-cultural training simulator, we developed a believability metric aimed at capturing these differences in *perceived* politeness.

In work funded by DARPA under the Force Multipliers for Urban Operations program, we completed an implementation and demonstration of an algorithm based on Brown and Levinson's model. We developed a "believability metric" for assessing and predicting when a given level of etiquette behavior exhibited by a computerized avatar or nonplayer character (NPC) in a game or simulation would be perceived as "unbelievable" by a human player or observer. As stated above (see Equation 4.2), Brown and Levinson theorized that the degree of face threat of any FTA must roughly equal the amount of redress used in order to maintain the social status quo. If less redress is used than is perceived as necessary—that is, if $W_x >> V(A_x)$—the utterance will be perceived as rude, and the hearer may seek alternative explanations or interpretations for the behaviors. If more politeness behaviors are used by the speaker than are perceived as necessary by the listener—that is, if $W_x << V(A_x)$—then the utterance will be perceived as over-polite or obsequious, and the hearer may look for ulterior motives or alternative interpretations of the situation. Thus, perceived politeness is a function of the imbalance (I) between the degree of face threat in an interaction and the degree of redress used. We can express this as:

$$I_x = V(A_x) - W_x \quad (4.3)$$

where I_x is the perceived imbalance of interaction x.

This equation states that the imbalance of interaction x will be the difference between the value of the redressive act A_x and the degree of face threat imposed, W. Thus, I_x will be positive when the speaker uses *more* redress than is warranted by the face threat posed by his or her utterance. A positive value of I_x corresponds to an overly polite interaction. I_x will be negative when the speaker uses less redress than the face threat warrants. A negative value of I_x corresponds to a rude utterance.

Note that there can be a great range in the *perceived* severity of the face threat posed in an interaction. An identical set of behaviors can be perceived as anything from rude to appropriate, to overly-polite, when used in different contexts or in different cultures. In other words, the same redressive value ($V(A_x)$) may be viewed as too little, just right, or too much, depending on the value of the face threat present, which, in turn, is dependent on the context of the interaction. Suppose someone says, "Pick up your books!" This may not be considered inappropriate when spoken by a parent to a child, but it will certainly be considered rude if spoken by the same adult to her boss.

Differences in perceived politeness may also occur between different observers of the same conversation because they may hold different beliefs about the power difference or social distance between the people interacting. Imagine two observers watching an interaction between a speaker and hearer. One observer knows that the speaker and hearer are close friends, while the other believes they are strangers. When the speaker says to the hearer "Give me five dollars," the first observer may view the speaker's request as not out of the ordinary, while the second one may view it as aggressively rude. This example illustrates that the same interaction could result in multiple values of W_x, each of which are correct from each individual observer's point of view. We can represent this as:

$$Bo{:}I_x = Bo{:}V(A_x) - Bo{:}W_x \tag{4.4}$$

where Bo is the Belief of Observer, O, about the other terms in the equation. Thus, this equation represents the imbalance from O's perspective. The ability to capture multiple interpretations of the same interaction is important in simulations involving multiple agents.

4.8 The Etiquette Engine in Cross-Cultural Training

We demonstrated the use of this method in the context of language training and mission rehearsal systems. Our politeness algorithm, which we have labeled the Etiquette Engine® (Miller et al., 2005; 2006; 2007), enabled nonplayer characters (NPCs, e.g., simulated computer-generated characters) to both recognize the degree of politeness directed at them and to reason about the level of politeness to be used in an interaction they themselves issue in keeping with other goals the character may have. We briefly describe examples of our implementations below.

4.8.1 The Tactical Language Training System

Our first software implementation of the Etiquette Engine has been in the context of the Tactical Language Training System (TLTS) developed by the University of Southern California's CARTE Laboratory (Johnson, Vilhjalmsson, and Santani, 2005). TLTS is a simulation designed to teach soldiers "tactical" versions of foreign languages. A trainee must navigate scenarios and interact with simulated local inhabitants (e.g., an Afgani store merchant) to accomplish an overall mission (e.g., negotiate a meeting time with a village elder). The outcome of the scenario depends on the trainee's verbal language skills as well as politeness skills in words and gestures. Prior to our involvement, TLTS used either a traditional scripting approach or a complex, first-principles reasoning engine called PsychSim to model human goal-based reasoning and decision making (Pynadath and Marsella, 2005).

A traditional scripting approach for coding NPC behaviors is very labor intensive. It required a programmer to predefine all possible interactions between an NPC and a user as a strict set of rules. However, it resulted in scenarios that were limited in scope and brittle; if the user performed actions that were not scripted *a priori*, then the NPCs would not understand the player's actions, nor respond with appropriate behaviors. PsychSim, a detailed goal-based reasoning module, increased the flexibility of TLTS to an extent but also increased the cost and computational complexity. PsychSim offers a rich and deep representation of human goal-based reasoning and includes similarly deep reasoning about the intentions of other agents.

While the complex model built with PsychSim was far richer than the Etiquette Engine that we later proposed, it required substantial development effort and produced comparatively unpredictable results. By using the Etiquette Engine, we were able to substantially reduce engineering efforts by eliminating the need to define all NPC behaviors *a priori*, and reduce the brittleness at the same time.

Brown and Levinson provide a very general and broadly applicable model of etiquette, but it lacks detail. Our refinements in the form of the Believablity Metric make it possible to represent individual interpretations of etiquette as well as those typical of specific cultural groups. This allowed us to incorporate "cultural modules"* in TLTS. A *cultural model* is a knowledge base of culture-specific interpretations of face threats and culturally appropriate sets of etiquette behaviors to redress them. The cultural model is used to control the perceptions and actions of the NPCs in the TLTS. The ability to represent alternate cultural perspectives by swapping cultural modules gave our algorithm a scalability that far exceeds traditional scripting approaches.

4.8.2 Interactive Phrasebook

In another project, sponsored by the Office of the Secretary of Defense, we applied the Etiquette Engine algorithm to a hand-held cross-cultural mission rehearsal tool called Interactive Phrasebook. The objectives of this tool were to familiarize the user with cultural knowledge and to help him/her reason about how social interaction "moves" will be perceived by people from various cultures.

As a demonstration, we developed a scenario template consisting of approximately 30 text interchanges in modern standard Arabic, in both formal and colloquial language. With the help of cultural experts we instantiated Interactive Phrasebook with a culture-specific knowledge base of phrases, redressive gestures, and power and familiarity markers (e.g., ethnic culture, occupation). A programmer can quickly configure various attributes to create customized scenarios, including the cultural background of each NPC, his or her general character attributes (i.e., tendency to be polite or rude), the power distance between NPCs, and their familiarity with each other, and the degree of face threat imposed by the topic of discussion.

Users can interact with the Phrasebook by choosing utterances and gestures for the character they control, and observing how the other characters react in the context of a discussion. In this way, users can explore the impact of various behaviors in a cultural setting very different from their own, and their effect on the outcome of the exchange.

---

* This work was conducted under a project funded by DARPA called "Force Multipliers for Urban Operations."

For example, did the negotiations succeed, or did they break down into dangerous hostilities? The prototype was created for the PDA form factor and can be downloaded to most portable devices running the Microsoft Windows CE operating system.

4.9 Challenges and Future Work

The Etiquette Engine's use of Brown and Levinson's model and theory to inform about social interaction behaviors guarantees (insofar as Brown and Levinson's work is correct) that it will be universal in its reasoning about and scoring of abstract politeness "moves." However, as Meier (1995) notes, while the concept of politeness exists in every society, there are no assurances that the means of communicating politeness, i.e., redressive strategies, are functionally equivalent across languages and cultures. In our computational model, we have introduced a means of weighting each component individually as a possible rudimentary means to compensate for the Anglo-centric nature of Brown and Levinson's model. Another area of study relates to the changes of perception after an interchange has occurred, and how they can be represented in our model. As an example, consider the tendency for social distance to decrease as the number of "good" interactions increase. The history of interactions can affect perceived power distance, social distance, or character, and therefore alter the subsequent amount of perceived face threat. Integrating such effects into the behaviors of NPCs will undoubtedly add to the simulation's realism. Further, Brown and Levinson themselves do not operationalize the parameters in their model; instead, they are offered as qualitative constructs. Evaluating content for use in the model is difficult because perceptions and interpretations of interactions are highly contextual. The model designers must tease apart the contextual information and assign an importance to each of them while keeping the model abstract enough so it can represent a large number of scenarios. Currently, our approach to content capture is a manual process that is labor intensive. A tool that can capture information from multiple cultural experts and transfer it into a computer-useable format will enable rapid development of multiple cultural modules.

4.10 Conclusion

We argue that models of etiquette can inform both human–human and human–machine interactions and perceptions. We have presented Brown and Levinson's theoretical model of politeness as an example of a model for human–human interaction, and have demonstrated that embedding etiquette rules in a machine can result in a number of benefits. By creating and making use of a formal computational model of etiquette, we have seen substantial savings in engineering time in the creation of simulations. More importantly, a computational etiquette model can enhance the user experience by creating a shared model of context, utilizing the richness available in human–human interactions to refine and increase the amount of useful information exchanged, and ultimately increase human + machine performance. We have already begun to see evidence of this in our own studies as well as the work of others (Parasuraman and Miller, 2004).

Advances in robotics, artificial intelligence, and computing power are working to materialize the concept of socially aware machines (e.g., Rickel and Johnson, 1998; Breazeal, 2002; Cassell and Bickmore, 2002; Kidd et al., 2006). However, socially believable machines will not be possible until we can represent, quantify, and assess the richness of social interactions in computational methods. We claim that etiquette is one aspect of social interaction that has practical implications in both human–human and human–computer interactions. We have demonstrated the feasibility of implementing a model of etiquette in a computational manner for several applications. By leveraging the large volume of work in sociology and linguistics, we believe that one can utilize an etiquette model's capability to reason about and significantly enhance human–machine and computer–mediated interactions.

References

Ardissono, L., Boella, G., and Lesmo, L. 1999. Politeness and Speech Acts. In *Proceedings of Workshop on Attitude, Personality and Emotions in User-Adapted Interaction*, pages 41–55, Banff, Canada.

Bates, J. 1994. The Role of Emotion in Believable Agents, *Communications of the ACM*, Special Issue on Agents, July 1994.

Breazeal, C. 2002. *Designing Sociable Robots*. Cambridge, MA: MIT Press.
Brown, P. and Levinson, S. 1987. *Politeness: Some Universals in Language Usage*. Cambridge Univ. Press; Cambridge, U.K.
Cassell, J. and Bickmore, T. 2002. Negotiated collusion: Modeling social language and its relationship effects in intelligent agents. *User Modeling and User-Adapted Interaction*. 13(1): 89–132.
Cialdini, R.B. 1993. *Influence: Science and Practice* (3rd ed.). New York: Harper Collins.
Cummings, M.L. 2003. Tactical tomahawk interface for monitoring and retargeting (TTIMR) experiment and simulation results. Final report to Naval Surface Warfare Center, Dahlgren Division.
Dorfman, P.W. and Howell, J.P. 1988. Dimensions of national culture and effective leadership patterns: Hofstede revisited. In Farmer, R.N. and McGoun, E.G., *Advances in International Comparative Management*, Vol. 3. Greenwich, CT: JAI Press.
Eelen, G. (2001): *A Critique of Politeness Theories*. Manchester: St. Jerome.
Goffman, E. 1967. *Interactional Ritual*. Chicago: Aldine.
Grice, P. 1975. Logic and conversation. In P. Cole and J.L. Morgan (eds.), *Syntax and Semantics*, Vol. 3, Speech Acts, New York: Academic Press, 1975.
Hofstede, G. 2001. *Culture's Consequences, Comparing Values, Behaviors, Institutions, and Organizations Across Nations*. Thousand Oaks CA: Sage Publications.
House, J. (2005). Politeness in Germany: Politeness in Germany? In Leo Hickey and Miranda Stewart (eds.), *Politeness in Europe*. Clevedon: Multilingual Matters, 13–28.
Johnson, W.L., Vilhjalmsson, H., and Samtani, P. 2005. The tactical language training system (technical demonstration). The first conference on Artificial Intelligence and Interactive Digital Entertainment, June 1–3, Marina del Rey, CA.
Kidd, C.D., Taggart, W., and Turkle, S. 2006. A Sociable Robot to Encourage Social Interaction among the Elderly. *2006 IEEE International Conference on Robotics and Automation (ICRA)*. Orlando, FL, May 15–19, 2006.
Lee, J.D. and See, K.A. 2004. Trust in technology: Designing for appropriate reliance. *Human Factors*, 46(1): 50–80.
Meier, A.J. 1995. Passages of politeness. *Journal of Pragmatics* 24: 381–392.
Miller, C., Wu, P., Krichbaum, K., and Kiff, L. 2004. Automated elder home care: Long term adaptive aiding and support we can live with. In *Proceedings of the AAAI Spring Symposium on Interaction between Humans and Autonomous Systems over Extended Operation*, March 22–24, 2004. Stanford, Palo Alto, CA.
Miller, C., Chapman, M., Wu, P., and Johnson, L. 2005. The "etiquette quotient": An approach to believable social interaction behaviors. In *Proceedings of the 14th Conference on Behavior Representation in Modeling and Simulation (BRIMS),* May 16–19, 2005, Universal City, CA.
Miller, C., Wu, P., Funk, H., Wilson, P., and Johnson, W.L. 2006. A computational approach to etiquette and politeness: Initial test cases. In *Proceedings of 2006 BRIMS Conference*, May 15–18, 2006, Baltimore, MD.

Miller, C. 2003. The etiquette perspective on human-computer interaction. In *Proceedings of the 10th International Conference on Human-Computer Interaction*, June 22–27, 2003, Crete, Greece.

Miller, C. and Funk, H. 2001. Associates with etiquette: Meta-communication to make human-automation interaction more natural, productive and polite. In *Proceedings of the 8th European Conference on Cognitive Science Approaches to Process Control*, September 24-26, 2001; Munich. (Invited).

Miller, C., Wu, P., Funk, H., Johnson, L., and Vilhajalmsson, H. 2007. A Computational Approach to Etiquette and Politeness: An "Etiquette Engine®" for Cultural Interaction Training. Awarded "Best Paper." *In Proceedings of the 2007 Conference on Behavior Representation in Modeling and Simulation*, March 26-29, Norfolk, VA.

Mori, M. 1970. Bukimi no tani—The Uncanny Valley (K. F. MacDorman & T. Minato, Trans.). *Energy*, 7(4), 33–35. (Originally in Japanese.)

Nass C. 1996. *The Media Equation: How People Treat Computers, Televisions, and New Media Like Real People and Places*. New York: Cambridge University Press.

Norman, D. 2004. *Emotional Design: Why We Love (or Hate) Everyday Things*. New York: Basic Books.

Parasuraman, R. and Miller, C. 2004. Trust and etiquette in high-criticality automated systems. *Communications of the Association for Computing Machinery*, 47(4), 51–55.

Pautler, D. 1998. *A Computational Model of Social Perlocutions*, COLING/ACL, Montreal.

Pynadath, David V., and Marsella, Stacy C. 2005. PsychSim: Modeling Theory of Mind with Decision-Theoretic Agents. In *Proceedings of the International Joint Conference on Artificial Intelligence*, pp. 1181–1186, 2005.

Reeves, B. and Nass, C. 1996. *The Media Equation*. Cambridge University Press, Cambridge, U.K.

Rickel, J.W. and Johnson, W.L. 1998. STEVE: A pedagogical agent for virtual reality (video) in *Proceedings of the Second International Conference on Autonomous Agents,* May 10–13, 1998, Minneapolis, MN.

Schricker, B.C., Schricker, S.A., and Franceschini, R.W. 2001. Considerations for selective-fidelity simulation. In *Proceedings of SPIE—The International Society for Optical Engineering*, Vol. 4367, pp. 62–70, Enabling Technology for Simulation Science V, Alex F. Sistis and Dawn A. Trevisani, Eds.

Searle, J. 1969. *Speech Acts: An Essay in the Philosophy of Language*. Cambridge, U.K.: Cambridge University Press.

Shirado, T. and Isahara, H. 2001. Numerical model of the strategy for choosing polite expressions. In *2001 Computational Linguistics and Intelligent Text Processing,* 98–109.

Thomas, F. and Johnston, O. 1995. *The Illusion of Life: Disney Animation*. New York: Disney Editions.

5

THE ROLE OF POLITENESS IN INTERACTIVE EDUCATIONAL SOFTWARE FOR LANGUAGE TUTORING

W. LEWIS JOHNSON AND NING WANG

Contents

5.1 Introduction

In recent years there has been significant progress in the development of advanced computer-based learning environments that interact with learners, monitor their progress, and provide individualized guidance and feedback. These systems are successfully making the transition out of the research laboratory and into widespread instructional use. For example, the mathematics tutors developed by Carnegie Learning have been employed by nearly 500,000 students in over 1000 school districts across the United States. Foreign language and culture courses developed by Alelo have been used by tens of thousands of learners in the United States and other countries around the world.

The goal of such systems is to promote learning gains that far exceed what is typical of classroom instruction. Bloom (1984) presented studies of learning involving one-on-one interaction with tutors, which suggested that learners who receive individual tutoring perform two standard deviations (2σ) better on average than learners who receive classroom instruction. Since then, developers of intelligent tutoring systems and other intelligent learning environments have sought to demonstrate learning gains that approach what human tutors are able to achieve.

For such learning gains to be realized, it is essential that learning systems work effectively for extended periods of time. It is important, therefore, for the learning system to promote learner interest and motivation. If learners are not motivated to continue working with a learning system and cease using it, the learning system cannot be effective. This is the fate of many learning systems designed for self-study, including most language-learning software.

This chapter describes efforts to apply principles of politeness to intelligent learning environments, so that the software interacts with learners in accordance with social norms of human–human interaction. This can be accomplished by incorporating animated pedagogical agents into the learning system, i.e., animated characters that are capable of interacting with the learners in natural, socially appropriate ways. We hypothesize a *politeness effect* that proper adherence to principles of politeness theory can help promote learner interest and motivation, and this in turn can have an impact on learning effectiveness. We present results of laboratory studies that show that politeness can have a positive effect on learning outcomes. We then discuss how we have applied etiquette in the Tactical Iraqi™, one of a family of language and culture learning systems that are in widespread, extended use. These results help to establish whether the politeness effect generalizes across subject matter domains and user populations, whether it occurs in authentic learning contexts, and whether it has a persistent effect over time.

5.2 Background: Pedagogical Agents

The work described in this chapter developed out of earlier research on the Media Equation and pedagogical agents. Starting in the 1990s, researchers Byron Reeves and Clifford Nass and their colleagues began investigating how results from social science research can also apply to human–computer interaction research. They proposed the Media Equation (Reeves and Nass, 1996), which stipulates that people respond to media, including computer-based media, as they do to other people. They argued that designers of computer systems should take this similarity into account, and be prepared for the possibility that users will attribute human-like attributes such as personality to their designs. For example, their studies showed that people tend to respond positively to other people who flatter them, and so therefore designers of computer systems should generate messages that flatter their users.

Soon after Reeves and Nass began publishing their work on human–computer interaction, researchers in intelligent tutoring systems began to investigate how the Media Equation might apply to educational software. Johnson, Rickel, and Lester (2000) began to investigate whether animated pedagogical agents could exploit the Media Equation to improve learning effectiveness. The idea was that if users tend to attribute human-like characteristics to computer systems, this effect might be accentuated by an on-screen character that assumes a human-like form. Researchers in intelligent tutoring systems had for some time been deliberately attempting to emulate aspects of one-on-one tutorial dialog to achieve the kinds of learning gains that Bloom (1984) described. An animated persona could emulate additional aspects of human–human interaction, such as displays of emotion and empathy.

Lester et al. (1997) reported a study in which an animated pedagogical agent named Herman the Bug facilitated learning with an intelligent learning environment named Design-a-Plant. They posited a Persona Effect, that a lifelike animated agent that expresses effect could facilitate learning. Researchers conducted further studies to attempt to replicate the Persona Effect in other domains and understand it in further detail. The results of these studies have been mixed. André, Rist, and Müller (1998) demonstrated that an animated agent could help reduce the perceived difficulty of instructional material, and Bickmore (2003) reported that subjects liked an animated agent that responded socially to them, but these agents did not yield differences in learning gains. Further studies (Atkinson et al., 2005; Graesser, et al., 2003; Mayer et al., 2004; Mayer, Sobko, and Mautone, 2003; Moreno and Mayer, 2000, 2004) suggested that it was the voice of the animated agent that influenced learning, not the animated persona.

Although we and others continue to explore the role of animated agents in promoting learning (Johnson et al., 2004a; Rosenberg et al., 2007), we have come to conclude that the animated persona per se may not be the primary cause of the learning effects of animated pedagogical agents (Figure 5.1). Rather, an animated pedagogical agent may promote learning if it can emulate the kinds of social interaction tactics that human tutors employ to promote learning. Animated personas convey the impression that they are social actors, in accordance with the Media Equation, but the agent must then behave appropriately as social actors in order for the agents to be effective. In our work we have found Brown and Levinson (1987)'s politeness theory to be particularly relevant to the design of interaction tactics for learning systems.

Figure 5.1 Gina, the animated pedagogical agent from the Bright IDEAS psychosocial intervention. (From Johnson, W.L., LaBore, C., and Chiu, Y.-C. (2004a). A pedagogical agent for psychosocial intervention on a handheld computer. In *Proceedings of AAAI Fall Symposium on Health Dialog System*. 64–70. Menlo Park, California: AAAI Press.)

5.3 Politeness, Tutorial Interaction, and Motivation

Brown and Levinson argue that people in all cultures have *face wants*, which influence their interactions with other people. They have a desire for positive face (the desire to be approved of by others) and for negative face (the desire to be autonomous and unimpeded by others). As explained elsewhere in this volume (Wu and Miller, in press), many interactions between people, such as requests and instructions, are potentially face-threatening, and so people employ a range of redressive strategies to mitigate face threat.

Tutorial interactions involve significant amounts of face threat. When tutors give instructions to learners, or tell them the right way to solve problems, they potentially threaten the learners' desire for autonomy. When tutors tell learners what they did wrong, they potentially threaten learners' desire for approval. Because teachers are in a position of power and authority over learners, they might be expected to disregard possible face threats and give direct instructions and feedback. However, our observations of human tutors suggest that they are very sensitive to potential face threats and make extensive use of politeness strategies to promote and maintain learner face.

A human tutor helping a learner with the Virtual Factory Teaching System (Dessouky et al., 2001) (a simulation-based learning tool for industrial engineering) phrased advice and instructions so as to avoid telling the learner directly what to do. Examples include "You will probably want to look at the work centers" (instead of "Look at the work centers") and "Want to look at your capacity?" (instead of "Look at your capacity"). The tutor also employed strategies to redress threats to positive face. These included avoidance of direct criticism of learner mistakes, and emphasizing joint activity, to promote a sense of solidarity between the tutor and the learner. Examples include "So the methodology you used for product 1 probably wasn't that good" (Instead of "You used a bad methodology for product 1") and "So why don't we go back to the tutorial factory" (instead of "Why don't you go back to the tutorial factory").

We studied a human tutor in the language learning domain tutoring Arabic pronunciation, and also found instances of face threat mitigation strategies. The tutor made attempts to avoid giving direct instructions, such as "Could you say that again?" (instead of "Say that again"). The tutor also avoided direct criticism of learner mistakes, e.g., "I didn't hear the 'H' in the word" (instead of "You left the 'H' out of the word"), and used tactics to promote a sense of solidarity, for example, "Good, we're ready to move on" (instead of "Move on"). Unlike in the virtual factory domain, however, the Arabic tutor also employed tactics to promote a positive attitude on the part of the learner, counteracting negative affect (e.g., "Don't worry, you'll be able to say 'H' soon") and playfully exaggerating positive performance (e.g., "Are you a native speaker or what?!").

Although politeness theory describes tutorial interaction at the tactical level, the politeness strategies described by Brown and Levinson cover only one aspect of it: how human tutors avoid offending and causing learners to lose face. But good tutors also try to actively cater to learners' face desires for autonomy and approval, and seek to promote them. They thus employ a wider range of politeness strategies than those Brown and Levinson described in their inventory of redressive strategies. This is particularly common among the Arabic tutors that we observed; beginning learners of Arabic often have low self-confidence in their abilities, and so tutors actively seek to boost their self-confidence. This is part of tutors' overall objective of motivating learners and encouraging them to devote effort to learning.

Researchers in motivation in learning, for example, Lepper and Hodell (1989), have identified factors that promote learner motivation, including the so-called Four Cs: confidence, curiosity, challenge, and control. Learners who are self-confident and feel that they are in control of the learning experience are more likely to be motivated to learn. Learning experiences that appeal to learners' curiosity, and that have an optimal level of challenge, tend to be more motivating.

Lepper and Woolverton (2002) studied highly effective tutors in remedial mathematics education, and found that they employed motivational tactics in their tutorial dialogs that seek to promote and optimize the Four Cs. Note that there is a close correspondence between the face wants identified by Brown and Levinson and some of these motivational factors. Negative face is related to control, and so tactics that address learner negative face may also influence learner sense of control. Positive face is related to self-confidence; if learners have a sense that others approve of their performance, they are more likely to be more confident of their own performance, which will motivate them to persevere. Thus, the politeness strategies that we observed in our studies of human tutors can be viewed as part of the tactical repertoire that tutors can employ to promote learner motivation.

If the Media Equation is correct, then politeness strategies and related motivational tactics could also be employed in computer-based learning environments to promote learner motivation. This in turn should result in improved learning. We call this effect the Politeness Effect. If the Politeness Effect holds, it could provide a tool to educational technologists to narrow the 2σ gap identified by Bloom (1984).

However the Politeness Effect, if it exists, is not likely to apply identically to all learners in all learning environments. The human tutors that we observed in our studies did not employ the same pattern of politeness strategies and motivational tactics. This may partly be due to differences in domains and tasks. Arabic is considered by English speakers to be a difficult language, and many speakers have difficulty pronouncing it correctly at first. Arabic instructors therefore perceive a need to encourage learners and build their self-confidence. Also, in the area of pronunciation and language form, there is relatively little scope for learners to exercise their autonomy. They either speak the language correctly or they do not. Politeness strategies that focus on learner autonomy may therefore have limited effect. In contrast, the subjects working with the Virtual Factory Teaching System had substantial latitude in planning their factory operations. Although they sometimes lacked self-confidence at first, they quickly gained confidence as they became familiar with the task and the system.

We therefore conducted a series of investigations, in the context of intelligent learning environments that we were developing, to apply politeness strategies and see whether a Politeness Effect results. We studied the phenomenon in multiple domains, with multiple user populations, in order to determine when the Politeness Effect is most relevant and how to apply it.

5.4 Application #1: Virtual Factory Teaching System

Our first study, conducted in the summer of 2004, was an evaluation of politeness strategies in the context of the Virtual Factory Teaching System (VFTS) (Dessouky et al., 2001). Further detail about this study may be found in (Wang et al., 2008). The VFTS helps learners acquire an understanding of industrial engineering by setting up and managing simulated factories. We augmented the VFTS with a pedagogical agent that could utilize politeness strategies in providing guidance and feedback. An animation engine generated the nonverbal gestures for the agent (Shaw et al., 2004), and a dialog tactic generator generated the verbal tutorial interventions (Johnson et al., 2003). The tactic generator took as input the type of tutorial intervention to be performed, and the desired degree of positive and negative face threat redress, and automatically generated a tutorial tactic with the appropriate combination of positive and negative politeness. The tactic generator could use politeness strategies to mitigate the face threat inherent in various tutorial dialog moves, and use additional amounts of positive and negative politeness to influence the learner's sense of confidence and control.

This study was conducted as a semiautomated Wizard of Oz study, focusing on the differential effects of polite and direct tutorial interaction. The subjects would interact with an experimental interface including a graphical interface for the VFTS, the animated pedagogical agent, and an online tutorial. A camera placed above the display focused on the subject's face and tracked the subject's eyes to determine where the subject was focusing on the screen. During the learning session, subjects would read through the tutorial, and carry out activities described in the tutorial. If the subject had a question, he or she could ask the pedagogical agent, using a text chat interface. The system then interpreted the subject's actions and presented its analysis to an experimenter, sitting in another room. It tracked the subject's focus of attention, based on the subject's focus of gaze and the location of the mouse on the screen. It tracked the subject's progress through the exercise, based on the sections of the tutorial that the subject was reading and the actions that the subject was performing on the VFTS. The system then presented its analysis to a human tutor in another room, along with the questions that the subject asked. The human tutor would then select high-level tutorial actions to perform, either in response to a question from the subject or because the subject appeared to be stuck or confused. The pedagogical agent then enacted the tutorial actions, automatically selecting tutorial tactics to employ to realize the tutorial actions selected by the human tutor. The system

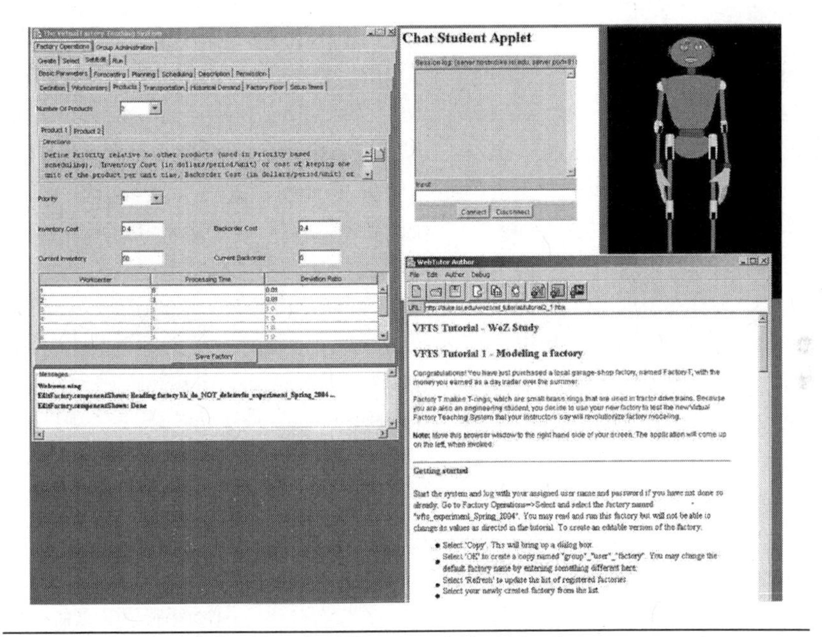

Figure 5.2 Learner interface for the VFTS pedagogical agent system.

was capable both of reactive tutorial responses and proactive tutorial interventions (Figure 5.2).

The study was organized as a Wizard of Oz study so that we could focus specifically on the role of politeness in the pedagogical agent's interactions with the subjects. The subjects were divided into two groups, a polite group and a direct group. For the polite group, the tutorial tactic generator was configured to generate polite tactics, based on levels of social power, distance, and desired influence on subject positive and negative face. The politeness values varied randomly in a range from moderate to high politeness, generating a variety of tutorial messages that would be generally appropriate in terms of level of politeness. For the direct group, the power, distance, and influence parameters were set to zero, causing the tutorial tactic generator always to generate tactics in the most direct manner possible. Further details regarding the tutorial tactic generator and its methods for selecting politeness strategies may be found in Johnson et al. (2004b). The tutor was experienced with the VFTS, and was asked to perform tutorial interactions as appropriate to the situation.

During the study, the subjects first completed some questionnaires measuring background knowledge and personality. An experimenter then provided the subject with a brief introduction to the system. The subject would then solve a basic assembly-line design problem with the VFTS, for approximately 35 minutes. The subject then completed post-test questionnaires measuring motivation and learning outcomes.

There were 51 subjects participating in the study: 17 students from the University of Southern California, and 34 students from the University of California–Santa Barbara. Among the student participants from USC, five participated in a pilot study, which allowed us to test the experimental set-up. Some subjects were excluded from analysis due to technical difficulties or because they had extremely high ability to perform the task without tutorial assistance. In the end, data from 37 students were analyzed. There were 20 students in the polite group and 17 in the direct group.

Overall, students who received the polite treatment scored better ($M_{polite} = 19.45$, $SD_{polite} = 5.61$) than students who received the direct treatment ($M_{direct} = 15.65$, $SD_{direct} = 5.15$). The difference is statistically significant, $t(35) = 2.135$, $p = 0.040$, $d = 0.73$. The scores ranged from 7 to 31 with a mean of 17.70. We analyzed individual characteristics of the subjects, including technical background and computer experience and personality. We found that the greatest individual differences were among subjects who preferred indirect help. That is, in the pretest questionnaires we asked the subject directly: did they prefer to receive instructional feedback in a direct manner, or in a polite, sensitive manner? For the subjects who preferred indirect feedback, the differences in learning outcomes were very pronounced, ($M_{polite} = 20.43$, $SD_{polite} = 5.74$, $M_{direct} = 13.00$, $SD_{direct} = 4.56$, $t(11) = 2.550$, $p = 0.027$). For the subjects who preferred direct feedback, there was much less of a difference.

We examined the subjects' attitudes toward the pedagogical agent. Did they like the agent, and would they want to work with the agent again? Subjects expressed a range of attitudes toward the agent: of the 37 subjects that we analyzed, 20 liked the agent and 17 either did not like it or had no preference. Of the ones who reported that they liked the tutor, the subjects who worked with the polite tutor performed better than the subjects who worked with the direct tutor ($M_{polite} = 20.33$, $SD_{polite} = 5.26$, $M_{direct} = 15.50$, $SD_{direct} = 4.96$, $t(18) = 2.058$, $p = 0.054$), especially on learning difficult concepts ($M_{polite} = 11.08$, $SD_{polite} = 2.61$, $M_{direct} = 8.38$, $SD_{direct} = 1.77$, $t(18) = 2.559$, $p = 0.020$).

There were 22 subjects who indicated that they would like to work with the tutor again, and among these, the subjects who worked with the polite tutor performed better on learning difficult concepts than subjects who worked with the direct tutor ($M_{polite} = 10.92$, $SD_{polite} = 2.75$, $M_{direct} = 8.50$, $SD_{direct} = 1.51$, $t(20) = 2.482$, $p = 0.022$). These results suggest that politeness strategies might have been beneficial in encouraging some learners to continue to learn; however, not all learners respond in the same way.

We attempted to measure learner motivation, using self-report assessments, to see whether the polite agent had an effect on learner motivation. We did not find a significant effect. However, it is difficult to draw conclusions from this, since the UCSB subjects were part of a psychology subject pool, had no choice of whether or not to participate in the study, and may have had little interest in the subject matter.

These results suggest that politeness may have a role to play in promoting the effectiveness of learning systems, and positive attitudes toward learning systems. Unlike many previous pedagogical agent studies that primarily presented results concerning subject attitudes toward the software, these results demonstrate an impact on learning outcomes, particularly for learners who express a preference for polite feedback. However, it is difficult to draw general conclusions from this limited laboratory experiment. The subjects used the VFTS for a limited amount of time, had no choice about whether or not to use it in the experiment, and did not necessarily have much interest in learning industrial engineering. Further investigation was needed in other domains, in authentic learning settings.

5.5 Application #2: Tactical Iraqi

To gain a better understanding of the role of politeness tactics and their effectiveness, we investigated their use in the context of another intelligent learning environment for learning Iraqi Arabic language and culture, Tactical Iraqi™ (Johnson, 2007). Tactical Iraqi is one of several game-based courses developed by Alelo, based on earlier prototypes developed at the University of Southern California. These courses help military service members who may have no knowledge of foreign language and culture quickly acquire functional communication skills. Others in current use are Tactical French™, focusing on the French spoken in sub-Saharan Africa, Tactical Pashto™, which teaches Pashto language and Afghan culture, and Tactical Dari™, focusing on Dari language and Afghan culture. Another course, Operational Indonesian™, is under development. An earlier prototype, Tactical Levantine™, was also developed, which teaches the Levantine Arabic dialect. Versions of these courses are available for the U.S. Army and Marines, as well as for military services in other countries. Alelo and its partners are applying the same instructional model in nonmilitary games that teach English, Chinese, Arabic, French, Cherokee, and other languages. Tens of thousands of people to date have learned foreign languages with these courses.

Figure 5.3 shows screenshots from Tactical Language courses, and examples of their use. They incorporate a Skill Builder, consisting of interactive lessons and exercises, and interactive game experiences. Learners use headset microphones to interact with the software, along with a keyboard and mouse. Lessons, exercises, and game experiences all involve speaking in the target language; speech recognition software is used to interpret the learner's speech. Learners also learn

Tactical Language learning lab installed for the U.S. Army 3rd Infantry Division at Ft. Stewart, GA.

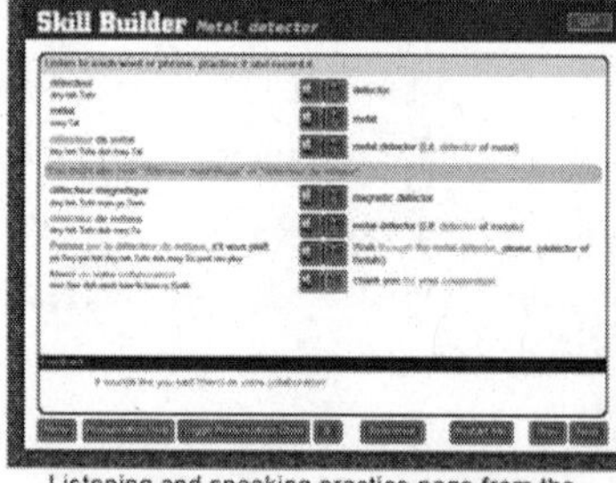

Listening and speaking practice page from the Skill Builder component of Tactical French 1.0

Figure 5.3 Images from Tactical Language and Culture training systems.

nonverbal gestures common to the target culture, and practice them in simulated face-to-face communication, selecting them using the keyboard or mouse.

Learners are introduced to concepts of language and culture in the Skill Builder lessons, and practice and apply them in the games. There are two types of game experiences. The Arcade Game is an example of a casual game, an engaging game experience that learners can play in their nonstudy moments, and which encourages repeated practice of selected communication skills. For example, in the game level shown in Figure 5.3 the learner gives spoken instructions to his character to navigate through a stylized town, collecting points along the way. The Mission Game gives learners opportunity to practice their communication skills in realistic scenarios. Learners move their characters through a 3D game world, speaking with a variety of non-player characters. To play the game successfully one must adhere to social norms of politeness and etiquette when interacting with the characters in the game.

Politeness and etiquette play multiple roles in Tactical Iraqi and related courses. The courses explicitly teach social norms of face-to-face interaction, via cultural notes and lesson pages. The lessons and game scenarios incorporate interactive nonplayer characters that respond to learners in accordance with principles of politeness. These provide learners with concrete opportunities to practice culturally appropriate face-to-face communication. The responses of the characters in the Mission Game provide learners feedback on whether they are communicating and behaving in a culturally appropriate way. Learners can also receive tutorial feedback and advice. The Skill Builder gives learners feedback on each spoken input and exercise response, and this feedback can be made to obey principles of polite tutorial dialog, as will be described in further detail below. When learners enter the Mission Game scenes they can be accompanied by a virtual aide character, who can make suggestions of what actions to perform and what to say. These suggestions also follow principles of politeness.

The Skill Builder gives learners feedback on their use of language; this feedback could in principle be delivered either in a direct way or in a polite way. The same is true for the advice given by the aide in the Mission Game. We therefore undertook some experiments to vary the use of politeness tactics in the tutorial interactions in Tactical Iraqi, and investigate whether this has an effect on learner motivation and learning effectiveness.

5.5.1 Incorporating Polite Tutorial Feedback into Tactical Iraqi

We focused our experimental manipulations of tutorial interaction on the Skill Builder, in order to produce the maximum experimental effect size. Feedback in the Skill Builder primarily is tutorial in nature, in contrast to the Mission Game where feedback comes primarily through interaction with nonplayer characters. Changing tutorial tactics would thus have little or not effect on the Mission Game.

To study how to incorporate tutorial feedback into Tactical Iraqi, we videotaped tutoring sessions with Tactical Iraqi. During the tutoring session, a professional teacher sat beside a student and was free to interrupt the learner while he/she worked with the program. Analysis of the videos revealed four different types of tutoring feedback:

- *Acknowledge/criticize*: Acknowledge that the learner action is correct or incorrect, and if it is incorrect, point out the errors in learner's action.
- *Elaborate*: Explain a language fact that relates to the learner's action. The tutor often uses various strategies to help students memorize the language fact.
- *Suggest action*: Offer hints as to what to do, for example "How about if you say /ma-ftihemit/?"
- *Recast*: Instead of explicitly criticizing the learner's action, the tutor simply demonstrates the correct action.

From the video, we also noticed that tutor would employ strategies purely for motivational purposes—for example, encouraging effort and consoling the student if they were having difficulty with particularly hard sounds.

Based on these observations we focused on implementing *acknowledge/criticize, elaborate, suggest action,* and *recast* feedback. We employed these in two types of Skill Builder pages: vocabulary pages and utterance formation pages.

User interaction on a vocabulary page consists of listening to the tutor's phrase, recording learner's own speech, playing back the learner's speech, and receiving feedback. The learner's speech is sent to an automated speech recognizer. The feedback model receives the recognized phrase from the speech recognizer and compares it to tutor's phrase (the correct phrase). If the learner's phrase matches the tutor's phrase, then the judgment of learner action is "correct," otherwise it's "incorrect." The second component of the feedback–learner action displays the phrase recognized by the speech recognizer, for example, "It sounds like you said 'as-salaamu aleykum [hello].'" The third component of the feedback offers the learner a suggestion on what to do next: practicing the utterance more, listening to the tutor speech, or moving on for now. The suggestion is selected at random from these three types. An example of the complete feedback could be "Your pronunciation is incorrect. It sounds like you said 'as-salaamu aleykum.' Try again." A pronunciation error detection model (Sethy et al., 2005) detected learner pronunciation errors. The feedback model could use its output to elaborate on pronunciation mistakes and construct more sophisticated suggestions.

Utterance formation pages require the learner to form a response to some prompt in some context. For example, a page might show a picture of an Iraqi man, along with the instruction: "Greet this man." The learner's speech is matched against a set of possible correct and incorrect responses. The feedback model then provides feedback depending on the response.

To apply the politeness strategies to the feedback, a database containing phrase templates using different politeness strategies was created. The feedback model queries the database with feedback component type and value, for example, "Judgment_of_action" and "correct." The database finds all the matches, selects one at random, and returns it to the feedback module (e.g., "Great job!") The feedback model combines the query results and delivers the feedback to the learner by an agent (Figure 5.4). Note that the agent is not animated and no synthesized speech is used. The feedback is delivered in text. For comparison purposes we also created a direct version, which simply indicates whether the learner's response is correct or incorrect (see Table 5.1).

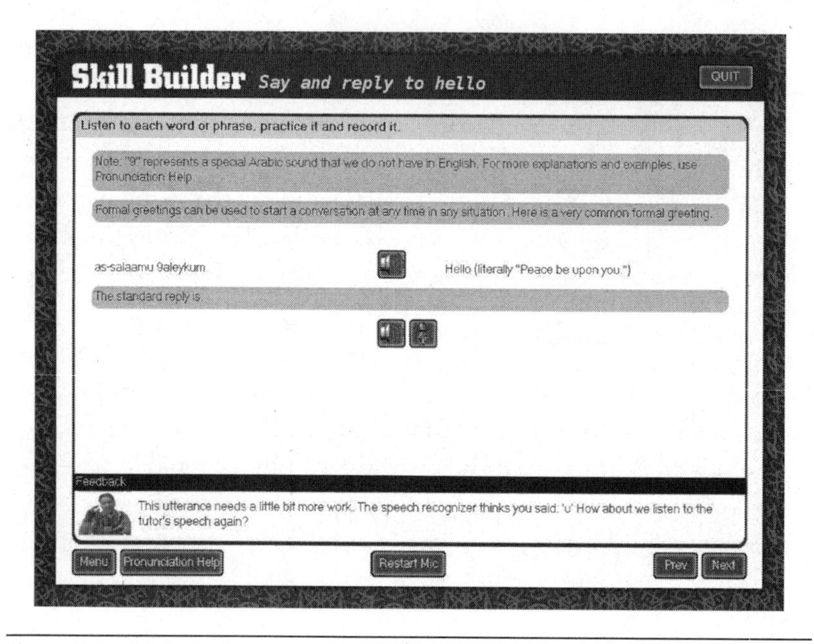

Figure 5.4 Feedback with appropriate politeness strategies delivered by an agent.

Table 5.1 Examples of Politeness Strategies

POLITENESS STRATEGY	EXAMPLE
Bald on record	Incorrect. Try again.
Exaggerate	Great job!
Common ground	Let's practice this a little bit more before we move on.
Be vague	Looks like someone got the right answer!
Understate	This utterance needs a little bit more work.
Question	How about we listen to the tutor's speech again?
Tautology	Practice makes perfect.
Impersonalize	It might be helpful to listen to the tutor's speech again.
Conventionally Indirect	This utterance requires more practice.

5.5.2 Evaluation of Feedback with Politeness Strategies

In order to conduct an evaluation with the maximum possible external validity, we needed to motivate users of Tactical Iraqi to use both the polite and direct versions. We therefore first released to our user community the version of Tactical Iraqi in which polite tutorial feedback was deliberately disabled, and then made available in a version with polite feedback at a selected test site, the Marine Corps training facility at Twentynine Palms, California. Both versions included a range of improvements over previous versions of Tactical Iraqi. This way, we avoided a situation where we were asking learners to use a version of Tactical Iraqi that was perceived as deficient or degraded. Had we not done this, trainers or trainees might have elected to use an older version of Tactical Iraqi and not use the experimental version.

5.5.2.1 Method

5.5.2.1.1 Participants Thirty-four Marines at Twentynine Palms participated in the study. The participants were male, from two different companies. We also collected log data from other Marines who were using Tactical Iraqi at the same training facility.

5.5.2.1.2 Design The study design was a between-subjects experiment with two conditions: Polite and Direct. Participants are assigned to two treatment groups randomly. We asked the units participating in the study not to use the game portions of Tactical Iraqi until they had completed at least two Skill Builder sessions, in order to maximize the effects of the tutorial feedback.

Learning gains were measured using quizzes at the end of each lesson. The quizzes contained utterance formation exercises, similar to the practice exercises described above, but with feedback turned off. Multiple choice and match exercises were also included.

Motivation, self-efficacy, and sense of autonomy were measured on both the pretraining questionnaire and the posttraining questionnaire with a few exceptions. There were three types of motivation measured in this study: language training motivation, motivation to transfer, and future training motivation (Mathieu, Tannenbaum, and Salas, 1992). These items are developed by SWA Consulting. Motivation to transfer and future training motivation were only measured posttraining. There are three types of self-efficacy measured in this study: learning self-efficacy, task-specific self-efficacy, and videogame self-efficacy (Bandura, 1995). Sense of autonomy was measured but in the posttraining questionnaire. Participants' performance in Tactical Iraqi was measured as an index of effort. These performance measures were obtained from data logged by the Tactical Iraqi system. Three items were obtained from the log: training time and listen-and-speak attempts in Skill Builder.

Within Tactical Iraqi, a pop-up survey was implemented to measure trainee motivation and affective state. The pop-up survey appears every 30 minutes. The survey contains two questions. The question on affective state asks trainees to rate themselves on a 1 to 7 scale: 1 represents “I am not enjoying learning Iraqi at all,” and 7 represents “I am really enjoying learning Iraqi a lot.” The question on motivational state asks trainees to rate themselves on a 1 to 7 scale: 1 represents “I really don’t think learning Iraqi is my thing,” and 7 represents “I feel like I am pretty good at Iraqi.”

Individual characteristics were measured in an attempt to address their interaction with the Politeness Effect. These individual characteristics include videogame experience (Orvis et al., 2006), experience with training technology (developed by SWA Consulting), extraversion (Donnellan, Oswald et al., 2006), openness (Goldberg, 1999), conscientiousness (Donnellan et al., 2006), preference for indirect help, attitudes toward language learning (Gardner, 1985), goal orientation (VandeWalle, 1997), and motivation instrumentality and valence (Vroom, 1964).

5.5.2.2 Result Out of the 34 Marines who participated in the study, only 23 completed both pre- and post-survey. Data from another 16 Marines were used as a control.

In terms of learning, the quiz scores of the polite group (M_{polite} = 6.30, SD_{polite} = 3.23) were lower than the quiz scores of the direct group (M_{direct} = 10.21, SD_{direct} = 5.91), contrary to our hypothesis. However the difference was not significant (p = .072). In terms of motivation, self-efficacy, sense of autonomy, and effort, we did not find significant difference between the two groups.

5.5.3 Discussion

Technical problems with speech-recognition-based pronunciation error detection limited the effectiveness of the feedback focusing on pronunciation errors. Pronunciation error detection in arbitrary spoken phrases is a difficult problem, and the error detection algorithm had difficulty performing it with confidence. If the recognition confidence was low, the feedback generator was forced to resort to vague messages praising effort, or qualified messages indicating that the learner's response *might* be correct. These could reduce the credibility of the feedback. This suggests that learners may react differently to computer-generated politeness strategies, depending upon their apparent purpose. They may respond positively to strategies that avoid causing the *learner* to lose face, but are less accepting of strategies that avoid causing the *computer* to lose face.

There are several critical differences between the feedback in VFTS and Tactical Iraqi. In Tactical Iraqi, there is no animated agent. The feedback is delivered through plain text instead of synthesized speech, which may reduce variability in the use of politeness strategies. This requires further investigation in future studies.

To further explore the relevance of politeness strategies in the language learning domain, we conducted a follow-up study in Tactical Iraqi with nonmilitary volunteers in the laboratory where the experiment is more controlled. We found that politeness strategies did make a significant difference in increasing learners' motivation and improve learning results (Wang and Johnson, 2008). This means that the military culture surrounding the subjects in the current study may have affected the effectiveness of the politeness strategies deployed, particularly the ones addressing negative face. The careful feedback to mitigate one's negative face, which is the need for autonomy, may clash with the unquestionably-follow-the-command military culture.

5.5.4 Implications for Tutorial Tactics in Tactical Language Courses

Based upon the experience with politeness strategies, we have adopted a practice of incorporating politeness strategies into the feedback messages employed in the Skill Builders of Tactical Language courses. Our authoring guidelines call for authors to write tutorial feedback that gives encouragement when learners give a correct answer, and which avoids being too critical when learners give an incorrect answer. We also check to make sure that diagnostic messages and other messages generated by the system avoid unnecessary threats to learner face. For example, if the speech recognizer is unable to recognize the learner's utterance, the system deflects blame from the system to the user, and displays a message "The system didn't understand what you said," instead of something like "You said that wrong."

However, we have found it necessary for various reasons to limit the role of polite tutorial dialog in Tactical Language courses. In earlier versions of the Mission Game, the learner's character was accompanied by an aide character that could offer hints and make suggestions. Because the aide could only offer one suggestion at a time, learners got the false impression that there was only one correct action in any given situation, and tended to rely on the aide to tell them what to do. We have since removed the aide role from the Mission Game, and provided learners with menus of possible actions to choose from.

Meanwhile, we have also been adapting the role of tutorial feedback in the Skill Builder. We have adopted a new method to providing pronunciation feedback, which is more reliable but does not emulate human tutorial dialog. Skill Builder exercises increasingly incorporate animated characters that the learner can speak to and respond to. This gives learners more practice in communicating with other people. It also reduces the amount of direct critical feedback the learner receives from the tutor. On the other hand, if the learner gives a correct response, then both the animated character and the tutorial feedback can give a positive response and promote positive face.

5.6 Conclusions

This chapter has presented experimental results measuring the Politeness Effect in multiple domains. Our studies also suggest that there are individual differences in the effect of politeness strategies, and there may be group differences in effectiveness as well. We have focused primarily on the application of the Politeness Effect in tutorial dialog; however, Tactical Iraqi demonstrates that there are alternative modes of interaction such as games that rely less on tutorial dialog, and the Politeness Effect may be less relevant to these alternative modes of interaction. Nevertheless, we continue to take the Politeness Effect into account in the design of tutorial feedback in Alelo language courses, where it is appropriate to do so.

Designers of educational software should consider carefully how politeness strategies apply to their particular application to avoid causing learners to lose face. In interactive applications that provide feedback, there are typically many opportunities to employ politeness strategies. Conversely, system developers and content authors who neglect politeness issues may unintentionally introduce messages that threaten learner face. On the other hand, learners may respond negatively to politeness strategies that they consider inappropriate, for example when the computer appears to be giving deliberately vague responses in order to avoid losing face. But overall politeness issues are an important aspect of the design of interfaces for educational software, and if addressed properly they can result in improved learner performance and improved learner attitudes and motivation.

References

André, E., Rist, T., and Müller, J. (1998). Guiding the user through dynamically generated hypermedia presentations with a life-like character. In *Proceedings of the 1998 International Conference on Intelligent User Interfaces*, 21–28. New York: ACM Press.

Atkinson, R.K., Mayer, R.E., and Merrill, M.M. (2005). Fostering social agency in multimedia learning: Examining the impact of an animated agent's voice. *Contemporary Educational Psychology, 30*, 117–139.

Bandura, A. (Ed.). (1995). *Self-Efficacy in Changing Societies.* New York: Cambridge University Press.

Bickmore, T. (2003). *Relational Agents: Effecting Change through Human-Computer Relationships*. Unpublished PhD thesis, Massachusetts Institute of Technology, Cambridge, Massachusetts.

Bloom, B.S. (1984). The 2 Sigma problem: The search for methods of group instruction as effective as one-to-one tutoring. *Educational Researcher*, Vol. 13, No. 6 (June–July, 1984), pp. 4–16.

Dessouky, M.M., Verma, S., Bailey, D., and Rickel, J. (2001). A methodology for developing a Web-based factory simulator for manufacturing education. *IEEE Transactions, 33*, 167–180.

Donnellan, M.B., Oswald, F.L., Baird, B.M., and Lucas, R.E. (2006). The mini-IPIP scales: Tiny-yet effective measures of the Big Five factors of personality. *Psychological Assessment*, *18*, 192–203.

Gardner, R.C. (1985). *Social Psychology and Second Language Learning: The Role of Attitudes and Motivation*. London: Edward Arnold Publishers.

Goldberg, L.R. (1999). International Personality Item Pool: A Scientific Collaboratory for the Development of Advanced Measures of Personality and Other Individual Differences [Online]. Available: http://ipip.ori.org/ipip/

Graesser, A.C., Moreno, K., Marineau, J., Adcock, A., Olney, A., and Person, N. (2003). AutoTutor improves deep learning of computer literacy: Is it the dialog or the talking head? In U. Hoppe, F. Verdejo, and J. Kay (Eds.), *Proceedings of International Conference on Artificial Intelligence in Education*, 47–54. Amsterdam: IOS Press.

Johnson, W.L. (2007). Serious use of a serious game for language learning. In R. Luckin et al. (Eds.), *Artificial Intelligence in Education*, 67–74. Amsterdam: IOS Press.

Johnson, W.L., Kole, S., Shaw, E., and Pain, H. (2003). Socially intelligent learner-agent interaction tactics. In *Proceedings of the International Conference on Artificial Intelligence in Education*, 431–433, Amsterdam, IOS Press.

Johnson, W.L., LaBore, C., and Chiu, Y.-C. (2004a). A pedagogical agent for psychosocial intervention on a handheld computer. In *Proceedings of AAAI Fall Symposium on Health Dialog Systems*, 64–70. Menlo Park, California: AAAI Press.

Johnson, W.L., Rickel, J.W., and Lester, J.C. (2000). Animated pedagogical agents: Face-to-face interaction in interactive learning environments. *International Journal of Artificial Intelligence in Education, 11*, 47–78.

Johnson, W.L., Rizzo, P., Bosma, W., Kole, S., Ghijsen, M., and Welbergen, H. (2004b). Generating socially appropriate tutorial dialog. In E. André, L. Dybkjær, W. Minker, and P. Heisterkam. (Eds.), *Proceedings of 2004 Affective Dialogue Systems, Tutorial and Research Workshop*, 254–264. Berlin: Springer.

Lepper, M.R. and Hodell, M. (1989). Intrinsic motivation in the classroom. In C. Ames and R. Ames (Eds), *Research on Motivation in Education 3*, 73–105. San Diego, CA: Academic Press.

Lepper, M.R. and Woolverton, M. (2002). The wisdom of practice: Lessons learned from the study of highly effective tutors. In J. Aronson (Ed.), *Improving Academic Achievement: Impact of Psychological Factors on Education*, 135–158. Orlando, FL: Academic Press.

Lester, J.C., Converse, S.A., Kahler, S.E., Barlow, S.T., Stone, B.A., and Bhogal, R.S. (1997). The persona effect: Affective impact of animated pedagogical agents. In S. Pemberton (Ed.), *Proceedings of 1997 Conference on Human Factors in Computing Systems*, 359–366. New York: ACM Press.

Mathieu, J.E., Tannenbaum, S.I., and Salas, E. (1992). Influences of individual and situational characteristics on measures of training effectiveness. *Academy of Management Journal, 4*, 828–847.

Mayer, R.E., Fennell, S., Farmer, L., and Campbell, J. (2004). A personalization effect in multimedia learning: Students learn better when words are in conversational style rather than formal style. *Journal of Educational Psychology, 96*, 389–395.

Mayer, R.E., Sobko, K., and Mautone, P.D. (2003). Social cues in multimedia learning: Role of speaker's voice. *Journal of Educational Psychology, 95*, 419–425.

Moreno, R. and Mayer, R.E. (2000). Meaningful design for meaningful learning: Applying cognitive theory to multimedia explanations. In *Proceedings of 2000 World Conference on Educational Multimedia Hypermedia and Telecommunications*, 747–752. Charlottesville, VA: AACE Press.

Moreno, R. and Mayer, R.E. (2004). Personalized messages that promote science learning in virtual environments. *Journal of Educational Psychology, 96,* 165–173.

Orvis, K.A., Belanich, J., and Horn, D.B. (2006). The impact of trainee characteristics on gamebased training success. In J.A. Cannon-Bowers (Chair), K.A. Orvis (Co-Chair), K.L. Orvis (Co-Chair), and S.J. Zaccaro (Discussant), *Learn N' Play: Effectiveness of Videogame-Based Simulations for Training and Development.* Symposium conducted at the 21st Annual Conference of the Society for Industrial/Organizational Psychology, Dallas, TX.

Reeves, B. and Nass, C. (1996). *The Media Equation.* New York: Cambridge University Press.

Rosenberg, R.B., Plant, E.A., Baylor, A.L., and Doerr, C.E. (2007). Changing attitudes and performance with computer-generated social models. In R. Luckin et al. (Eds.), *Artificial Intelligence in Education*, 51–58. Amsterdam: IOS Press.

Sethy, A., Mote, N., Narayanan, S., and Johnson, W.L. (2005). Modeling and automating detection of errors in Arabic language learner speech. In *Proceedings of the 9th European Conference on Speech Communication and Technology*, Lisbon: ISCA.

Shaw, E., LaBore, C., Chui, Y.-C., and Johnson, W.L. (2004). Animating 2D digital puppets with limited autonomy. In *Proceedings of the 2004 International Symposium on Smart Graphics*, 1–10. Berlin: Springer.

VandeWalle, D. (1997). Development and validation of a work domain goal orientation instrument. *Educational and Psychological Measurement*, 57, 995–1015.

Vroom, V. H. (1964). *Work and Motivation.* New York: John Wiley & Sons.

Wang, N., Johnson, W.L., Mayer, R.E., Rizzo, P., Shaw, E., and Collins, H. (2008). The politeness effect: pedagogical agents and learning outcomes. *International Journal of Human Computer Studies*, 66(2), 98–112.

Wang, N. and Johnson, W.L. (2008). The politeness effect in an intelligent foreign language tutoring system. In *Proceedings of the 9th International Conference on Intelligent Tutoring Systems,* 270–280. Berlin: Springer.

Wu, P. and Miller, C. (in press). Computational models of etiquette: Implementations and challenges. In C. Hayes and C. Miller (Eds.), *Human-Computer Etiquette.* London: Taylor & Francis.

6

Designing for Other Cultures

Learning Tools Design in the Nasa Amerindian Context

SANTIAGO RUANO RINCÓN, GILLES COPPIN, ANNABELLE BOUTET, FRANCK POIRIER, AND TULIO ROJAS CURIEUX

Contents

6.1 Introduction

This chapter describes our Cultural Model and its use in development of a computers tool targeted for a specific minority culture, a Colombian Amerindian community called Nasa (Paez). Many of the metaphors used in standard computer (Western)–computer interfaces have little relevance to their culture, rendering many applications conceptually inaccessible even when computers are available. Our cultural model was inspired by existing cultural models, which have different objectives from software design. We demonstrate how culture impacts both the interface design and the software development process. The interface design must incorporate the culture's conception of the world, symbols, and social norms for interaction. The software development process must also incorporate social norms in order to gain community acceptance. We will illustrate these points by recounting the design and development of a pedagogical game for use by Nasa school children.

The Southern Andes Colombian Mountains between el Cauca and Huila is the home of the Nasa people. The Nasa (Paez, as they are called in Castilian) are one of the Colombian native communities that are still alive. They are an agricultural people who consider themselves to be one of many beings in Nature. They are a nation with their own language, traditions, and *cosmovisión* (e.g., worldview).

As there are fewer and fewer frontiers in the world, personal computers and other modern technologies have arrived at every spot on the planet, and *Kiwe*, the Nasa territory, is no exception. Nasa people use PCs to write texts, listen to music, play games, etc. But computers are not yet fully understood by the Nasa people, at least not in the ways that software developers may expect.

In late 2005, there was an effort to translate basic computer terminology into Nasa Yuwe (the Nasa language) (Checa Hurtado and Ruano Rincón, 2006). Abelardo Ramos, Nasa linguist and leader, proposed a dozen terms, which were evaluated in a meeting by a group of Nasa students, teachers, and elders, all of whom were PC users or had basic knowledge about computers. The first turn focused on the terms *mjiwa' fxnu* and *mjinxi yũhne.* The group debated about them for some time and concluded that they meant a space for working. But when it was explained that they were suggestions for "desktop," people laughed off the idea saying, *"The desktop should be below the computer, not over it."* They were not conscious that such an element in the interface was called a desktop.

During the same process, common desktop icons were analyzed through a survey, and we identified several issues in the interpretations made by Nasa people. For example, we asked what kind of function they thought an icon like the clipboard (Figure 6.1) performs in the interface. This icon is usually used for the *paste* action, but none of the users could give an answer; they just left the space blank.

We asked them the same question for the recycling bin (Figure 6.2). These were some of their answers:

- "To receive"
- "When you do not want a file, you send it to this can called 'Recycle.'"
- "Garbage can to throw out paper"

At the same time, we asked them, *"What kind of object do you use to throw out the garbage?"* They answered:

- An old pot
- A plastic bag or can
- An old bucket or a big bag

Figure 6.1 Paste icon.

Figure 6.2 Recycling bin icon.

There were clearly some interpretation problems. The desktop analogy was not designed for the Nasa culture, thus, icons and general metaphor might be interpreted in unexpected ways.

The model presented here is an outcome and reflection of our work on the impact of cultural and anthropological factors in designing the interactions between computers and Nasa people.

We were facing the problem of how to develop and test computer tools for people from a culture different than the designer's. Nancy Hoft states that when designing international interfaces, a cultural model is required to understand users from each culture (Hoft, 1996). This is the approach that we have taken.

Nasa culture is in danger of disappearing. Their leaders and elders are trying to encourage its continuation. They promote clear policies to motivate people to speak their mother language and preserve their ancient traditions (Rappaport, 2005). Our overall goal in this work is to help these efforts, providing computer tools to be used in Nasa classrooms—tools that reflect Nasa culture.

We consider human–computer interaction to be a communication process influenced by norms, rules, and culture. This communication process refers to *etiquette* in the terms proposed by Hayes et al. (2002), asserting the adoption of content and mode of dialog between people, and by extension, between humans and computers in order to anticipate most common expectations in a given context.

User expectations in this process may impact their perception of technology and its use. So, to increase the likelihood that they will accept, adopt, use, and benefit from computing tools, we must respect their culture in our design.

6.2 The Cultural Model

We are not proposing a radically new cultural model. Our approach is based upon the work of Edward Hall (1989, 1990), David Victor (1992), Hofstede, G. and Hofstede, G.J. (2005), Fons Trompenaars (1993), and Kluckhohn and Strodtbeck (1961). Each of these cultural models was initially conceived with a specific goal, according to particular needs. Victor's LESCANT model helps in determining the main aspects of culture that affect business communication. Hofstede's model determines the patterns that form a culture's mental programming. Trompenaars' efforts are focused on determining the way in which a group of people solves problems.

In contrast, our work asserts that one needs to comprehend a culture in order to design effective interfaces for that population. We have compared our current experience and knowledge of the Nasa people with the variables presented by those existing models, and we will select and adapt the ones relevant to our goal. We aim to design human–computer interactions that fit with the cultural aspects and traditions of a given culture, while considering these interactions in the context of a large spectrum of activities. To achieve this, our cultural model borrows from many of the models above.

Under the premise that human–computer interaction is a communication process between the system developer and the user, we expect the cultural model to support a semiotic process in the interaction (de Souza, 2005), and to guide production of signs that are compatible with the user's understanding.

Relying on the analysis of main ethnographic and anthropological elements of Nasa culture, we have picked a subset of relevant "variables" presented by classical culture models and adapted the related dimensions to our goal. These dimensions are:

- Language
- Space
- Environment and technology
- Social organization
- Temporal notion
- Nonverbal signs

In the following paragraphs we present the main reasons and arguments for selecting these dimensions. In the following section, we will describe how the methodology was applied and validated as an example, and show how it is more generally relevant to design of computer tools.

6.2.1 Language

David Victor (1992) states that language is one of the most influential factors in culture. Relative to other cultural variables, language is easily visible, and most software localization processes take it into account, at least at a translation level. Characteristics of languages affect not only text in interfaces, but other features as well. Traditionally, elements on a PC screen are structured from left to right and top to bottom, as most European languages structure writing. Arabic systems change the left to right structure to right to left, following the convections of Arabic writing.

In our case, we have taken into account *Nasa Yuwe*, the Nasa language, in design of interfaces. It is spoken by more than 100,000 people, which constitutes around two thirds of the whole Nasa population. It was an oral tradition language until 2001, when a unified alphabet was accepted by the community (Rojas Curieux, 2002). It is a rich alphabet of 69 graphemes based on Latin, the outcome of a "basic" group of vowels and consonants and their transformations, produced by nasalization, palatalization, aspiration, and other phenomena. Complex characters are written with two or three letters—for example, ih, ẽe, çx, ph, txh, etc. (see Figure 6.3).

Language may carry a rich debate about neologisms and equivalence of meaning between languages. During the effort in 2005 described above, we tried to follow the localization process suggested for GTK+*. The same process has produced translations into German, French, Spanish, and dozens of other languages, but it was not successful for translating "basic" text strings such as "Yes," "No," "Color," "Open," etc., used everywhere in PC interfaces.

6.2.2 Space

This variable pertains to the ways in which cultures conceive, use, and structure space. It deals with undeclared rules pertaining to the layout of homes, land, and towns. Nancy Hoft (1996) takes the space concept in Hall's model and categorizes it into four sections: *Personal Space*, *Territoriality*, *Multisensory Space*, and *Unconscious Reactions to Spatial Differences*. We consider the first two of these categories, as we believe they are the most related to our goal.

Personal Space. Cultures have their own perceptions of personal space and follow unspoken and unconscious rules to determine where

* GTK+: A widely used toolkit to creating graphical user interfaces. http://www.gtk.org

	Oral Vowels				Nasal Vowels			
Short	a	e	i	u	ã	ẽ	ĩ	ũ
Glottal	a’	e’	i’	u’	ã’	ẽ’	ĩ’	ũ’
Aspirated	ah	eh	ih	uh	ãh	ẽh	ĩh	ũh
Long	aa	ee	ii	uu	ãa	ẽe	ĩi	ũu

Basic consonants	p	t	ç	k	m	n	b	d	z	g	l	s	j	y	w	r
Palatalized consonants	px	tx	çx	kx		nx	bx	dx	zx	gx	lx	sx	jx	yx	vx	
Voiceless aspirated stops consonants	ph	th	çh	kh												
Voiceless aspirated palatalised stop consonants	pxh	txh	çxh	kxh												

Figure 6.3 The Nasa Yuwe alphabet.

boundaries lie, and when these boundaries are violated. In Northern Europe, for example, people do not touch each other "and even brushing the overcoat sleeve of another in passing is enough to warrant an apology" (Hoft, 1996). Taking this factor into account will lead to different coding and representations of proximity, between people and entities, depending on culture.

Territoriality includes concepts about "ownership," and extends to how space communicates power. An example of this is the interior layout of offices. In some cultures, power or status of people is visible in the extension of space, while in others, there is no clear difference. In the U.S., people with power have large offices on the highest floors in buildings, but it is difficult to know who has authority in Japan, based on the same patterns.

Nasa culture exhibits many elements and traditions that may be attached to space and topological structures. In the Nasa context, the center of the family space is the *Tulpa* (Nasa hearth, built with three stones). The family shares and interacts around it. Seated around the fire, elders tell stories that everybody listens to carefully. That is one of the ways history has been transmitted in this oral tradition culture.

Beyond personal space, *Resguardos**, and the schools found on them are collective spaces which do not belong to a single person, but to the whole community. In the same way, they do not concentrate power in a few people, so layouts do not reflect the degree of authority.

Beyond these two elements inspired by Hoft's and Hall's approaches, we propose that a third element of space structuring that should be taken into account when designing an interface.

Space structuring. Cultures may have particular ways to conceive of and structure space. Nasa worlds, for example, are divided vertically in two sections: left and right. From a western point of view all kinds of elements that could be considered "positive," such as the sun, *Tays* (Energies), the *thẽ' wala* (Nasa traditional healer), etc., are found

* Communal and inalienable indigenous lands. The *resguardo* is the territorial unit that allows collective ownership over traditional lands by native communities. They are administered by elected councils (cabildos) and legitimized by colonial titles (Rappaport, 2005).

Figure 6.4 *Thẽ' wala* executing a cleaning ritual.

on the left. "Negative" elements, like the moon, disharmonies, cold plants, etc., are found on the right (Portela, 2008).

Energies move from one side to the other, causing disharmonies in energies and the cosmos. They can be restored through the symbolic cleaning ritual, where the *thẽ' wala* draws semicircles with medicinal plants, covering the body from the right to the left foot (see Figure 6.4) an image from the video "Nasa yuwe' walasa'—Nuestra lengua es importante" [CRIC, 1996a].)

The way space is structured is also important in the *Tul*, the Nasa house garden. There are rules that have to be respected to preserve harmony in Nature. Some plants must be placed on the left, others on the right; a big tree should not be located next to a small plant; a "hot" plant must not be placed beside a "cold" or a *rebelde*, etc. (CRIC, 1996b; Portela, 2008).

User interfaces may have an important spatial component. Space, a broad dimension in the Nasa case, includes personal, family, and common spaces. Or more accurately, different levels of social interaction, which start at the *tulpa* and may be described by a spiral (see Figure 6.5). It may be possible to take advantage of this notion to design and structure user interfaces. It determines, as well, how to locate "positive" and "negative" elements, event representations, and their dynamics in space.

To summarize, space structuring by the Nasa people emphasizes dimensions of positiveness and negativeness and also supports a

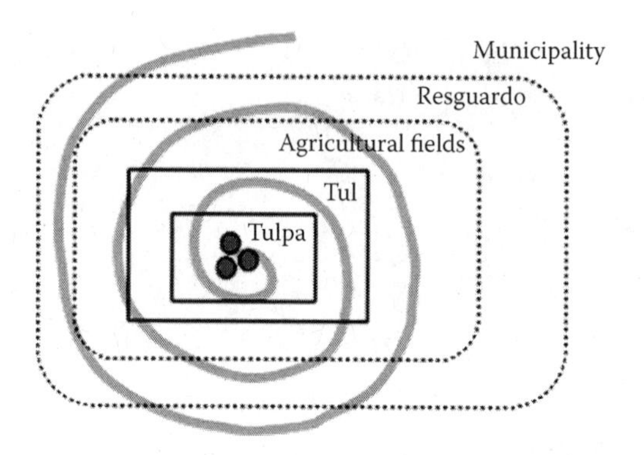

Figure 6.5 Social interaction levels.

concept of renewal realized by dynamic moves across the different regions of space. These dimensions are central in the perception of space, objects, and entities, and must be part of our cultural model.

6.2.3 Environment and Technology

This dimension studies the nature of the environment in which people live and work. It examines how geography, population, and perception of technology affect culture and how people adapt to their environment. David Victor (1992) states that people adapt to the natural world that surrounds them. In doing so, they define an essential part of the culture. According to Victor, environment provides the building blocks from which culture is constructed (Victor, 1992).

Most of the Nasa population inhabits *resguardos* in the Andes Mountains. There, they take advantage of the wide range of climates to cultivate several kinds of plants and vegetables: maize, manioc, coca, beans, coffee, sugar cane, potatoes, plantain, *fique* (sisal), some fruit trees, among others (Rojas Curieux, 2002). At the same time they try to live in harmony with Nature (Cátedra Nasa Unesco, 2001; CRIC, 1996b); they feel themselves to be a part of it, like any other being.

Agriculture is present even in the family environment. The *Tul*, an important cultural element, is found in every house. Besides being the house garden, it is, at the same time, a representation of the cosmos. It is also considered a technological model for agriculture and the main family sustenance.

Some plants have a special meaning. Maize, an American native plant, has been one of the main staples of the Nasa diet since ancestral times. It is never planted alone, but in conjunction with other species. The coca plant is essential in culture: it is used in daily life, and by the *thẽ' wala* in some rituals.

Offices are not typical in the Nasa environment. As we described earlier, the desktop analogy, commonly used in user interfaces, does not have the same meaning when used in Nasa culture. Even though there are Nasa people who use computers, we realized that the desktop metaphor does not resonate strongly with their experiences. Environment and technology also need to be considered in the physical design of an interface. For example, when computers have an outdoor use, as in an agricultural school environment, special consideration should be given to screen visibility, energy consumption, network access, etc.

6.2.4 Social Organization

In his LESCANT model, David Victor (1992) broadly defines social/organization as the common institutions and collective activities shared by members of a culture. Those institutions and activities influence people's behavior in all aspects of life, including communication processes, therefore, interaction with computers and machines.

Among the social structures that affect the culture considerably, we can find:

- Individualism and Collectivism (close to Hofstede's proposal)
- Conception of authority
- Family structure
- Educational system
- Class system
- Gender roles
- Religion
- Political system
- Mobility and geographical attachment
- Recreational institutions, etc.

Nasa society is characterized by a high level of collectivism. The *resguardos*, where most of the Nasa population live, are collective property. Each *resguardo* is governed by a *cabildo*, which are grouped in the *CRIC* (Regional Indigenous Council of Cauca). This is a political organization that represents indigenous communities in the national and regional sphere.

CRIC was born with the goal of recovering ancients lands, but has increased its area of responsibilities, including education in community schools, through the *PEBI* (Intercultural Bilingual Education Program).

The *PEBI* aspires to offer education in Nasa Yuwe and Castilian to Nasa children, providing a curriculum that follows community traditions and practices (Rappaport, 2005). It was designed as a way to preserve culture and strengthen Nasa politics, through school (CRIC and Rappaport, 2004).

The community is actively involved in pedagogical centers. The presence of Elders (called *mayores*) and *thẽ' walas* in the "class room" is frequent and welcomed, as they contribute to children's learning.

As in houses, the *Nasa tul* is an important element in schools. It is ubiquitous in the educational system and is used teaching and learning all subjects. Children learn math by measuring the different sizes of plants, and Castilian by writing texts about the garden*. Joanne Rappaport describes the *Tul* as a "microcosm in which *cosmovisión* can be researched, defined, and generated by local teachers and students" (Rappaport, 2005).

A typical activity of Nasa and other Amerindian cultures is the *minga*. Actually, this is the term used in Castilian by several native communities to describe a communal work party, while in Nasa Yuwe, people say *pi'kx yat* (the minga house). A *minga* is called when they need to build a school, carry out farming activities, celebrate a marriage (Cátedra Nasa Unesco, 2001], make a decision on a political issue, or any event where the community is involved (Wikipedia, 2009).

Abelardo Ramos states that the *minga* is an everyday practice. This concept was also used in the translation of the Political Constitution of 1991 into Nasa Yuwe, accomplished by an intercultural and

* Abelardo Ramos, personal communication, 2009.

interdisciplinary team (Rappaport, 2005). Like other Nasa practices, *minga* is also learned in schools.

Different levels of social interaction can be viewed from the spatial dimension. They start at the *Tulpa* (Nasa hearth), where the family meets and interacts, enclosed by the *tul*, which protects the house. They are surrounded by agricultural fields, and in turn, by the resguardo, the municipality, the department, and so on. This is represented by a spiral by Mincho Ramos, who is a Nasa leader, *PEBI* teacher, and Abelardo Ramos' brother (Rappaport, 2005).

As a highly collective society, Nasa people believe that information must not be controlled by a small group of people because this would represent a power concentration, which is not accepted. Thus, in schools computers do not have passwords or control access so that they may be used by everybody.

In evaluating interfaces, we have found that it has not produced meaningful outcomes to perform usability tests such as *card sorting* or verbal protocol methods *thinking aloud* with just one Nasa individual. We have obtained more successful evaluations through collective testing in group meetings during which everyone may speak and give his/her opinion.

We interpret this phenomenon as the expression of *etiquette* in Nasa behavior, and more precisely, in their social behavior. It implies that *etiquette* and culture apply not only to the content and dialog protocol of an interface but also to the design process itself, from the way in which the project is formulated, to validation. Retrospectively, this last assertion appears to be almost clear: validating one Human–Computer Interaction protocol or elementary component places local users in cultural situations where their behavioral expectations must be fulfilled.

6.2.5 Temporal Notions

Time is considered by all the cultural models described earlier, but with different perspectives. Hall (1990) classifies the time variable into two categories: *polychronic time* and *monochronic time*. They refer to two different ways of organizing and using time and space. This variable pertains to temporal issues and interrelated elements such as communication, human relationships, and activity development.

Polychronic time refers to the preference for doing many actions at the same time. It emphasizes involvement of people and completion of transactions, rather than preset schedules. Human relationships are more valuable than tasks or jobs. Examples of Polychronic people are Middle Eastern, Latin American, and Mediterranean cultures.

The other side, *monochronic time,* is characterized as "being sequential and linear." It stresses schedules, segmentation, and promptness. Monochronic people try to respect planned timetables as much as possible. This classification includes Northern European cultures. At present, we do not have enough information to classify Nasa culture into either of these categories.

From another perspective, Kluckholm and Strodtbeck (1961) propose a classification of time according to attitudes toward the past, the present, and the future.

Past-oriented cultures show respect for elders and ancestors. The origins of family, nation, or history in general are frequent subjects in conversations or public talks. The context of tradition or history is a common point of view.

Present-oriented people consider activities and enjoyment of the moment as the most important. Present relationships, here and now, have special interest for them. Such people do not object to future plans but they are rarely executed.

Future-oriented cultures are mostly interested in youth and future potential. They frequently talk about aspirations, prospects, and future achievement. "Planning and future strategy are regarded as very essential and done enthusiastically" (Lee, 2001).

The Nasa temporal conception has a strong orientation with the past. As described in Kluckholm and Strodtbeck's approach, elders and predecessors are highly respected, and their advice sought in all important decisions. This characteristic is common in oral-tradition cultures. Nasa people believe that ancestors show the way for current and future actions. Past is a component of the present and both past and present show the way to the future. Past, present, and future are distinguished but not processed separately by Nasa people. The relationship of events in history is not chronological, but logical. Additionally, the Nasa notion of time has a strong relationship with space, every single event having a link with the place where it happened (Valencia, 2000).

Temporal notions also influence how activities are developed when using the PC, how objects are represented and their evolution through time. For the Nasa, which is a *past-oriented culture*, elders, ancestors, and past events guide the way to the future. In keeping with this concept, we recommend that old documents should be made accessible as a base for further versions.

By extension of the politeness concept, their way of respecting elders, and consequently interacting within the community, is a very important part of our cultural and design model. Respecting culture and adhering to their etiquette has been proven to be the only effective way to introduce changes in the culture, especially technological changes, like the unified alphabet, accepted by the whole community.

6.2.6 Nonverbal Signs

Victor's (1992) model defines Nonverbal Behavior, from the business communication point of view, as "the exchange of information through nonlinguistic signs" (Harrison, 1974). This is a broad category that includes many types of human-produced signs and nonverbal behaviors. Although Victor's approach is focused on communication, we borrow his idea and restrict it to our work context.

According to Victor, examples of nonverbal actions are:

- Movement: A broad range of nonverbal activity, including facial expressions, waking rhythm, body posture, gestures of the hands, head, arms, etc.
- Colors: Describes preferences for and rejection of specific colors.
- Numbers and counting indicators: Describes the culture's number system.
- Symbols: Emblems, tokens, or other signs associated with the culture.
- Sounds: Associated with events, like incorrect interface manipulation, warnings, errors, etc.

Gestures, numbers, and colors are crucial to Nasa behaviors. Nasa movement of energies described in Section 6.2. can, for instance, be categorized here, as well as the clockwise spiral, the *ũhza yafx,* and other important symbols, described in the next section. Therefore, we consider nonverbal signs as an important part of our cultural model, and as complementary to more traditional "signs."

6.3 The Cultural Model in the Development Process of a Pedagogical Game

In this section, we describe how we applied the cultural model to the Nasa culture, present the methodology we used for development and testing of a school game, and discuss preliminary results of the validation. *Çut pwese'je*, "the maize game," is a word guessing game, which is similar to the popular *hangman*, but is designed to be used by people from the Nasa community with the aim of helping children to learn the Nasa Yuwe alphabet (Checa Hurtado and Ruano Rincón, 2006). This game was developed through an iterative process of several cycles. Changes between iterations were motivated, in most cases, by cultural characteristics that we found to have an impact on the expectations of the users. Cultural characteristics were identified through an ethnographic analysis, bibliographical research, observations during user testing, our prior knowledge of the Nasa culture, and interviews with Nasa people, linguists, and experts.

Note that expectations which users had for interactions with the computer were not necessarily expressed explicitly by users, as they are part of her/his culture. As we learned more about the Nasa culture, we understood better how they would interpret the interface and what they would expect from it.

6.3.1 Cycle Number 0: Initial Requirements

With the help of Abelardo Ramos and Tulio Rojas Curieux, the developers identified an initial set of requirements and game properties, which are characterized by the Cultural Model (see Figure 6.6).

6.3.1.1 Social Organization Nasa people are running campaigns to avoid losing their traditions and culture. Schools are oriented by the community through the Indigenous Regional Council of Cauca and the Bilingual Education Program. To coincide with their policies, the designers gave the game a pedagogical motivation: to help children to learn Nasa Yuwe.

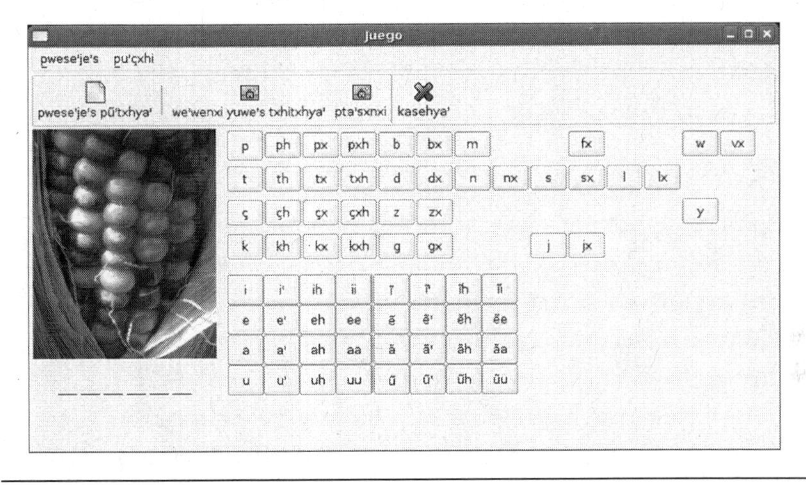

Figure 6.6 First interface prototype of *çut pwese'je.*

The first step in developing the game was to introduce the ideas and motivations to the community and get their approval. The community, as represented by teachers and leaders, was involved throughout the whole process, and they were always consulted about the game design. Nasa people are a collective society. For them, a group of people who control information also concentrate power, which is not accepted. Thus, the game requires no passwords, and anyone can make new entries in the word list used by the game.

6.3.1.2 Language Some of the Nasa Yuwe graphemes use two or three letters (e.g., ph, çx, txh), while others are not present in the Castilian alphabet (e.g., ĩ, ẽ, ã). To help children to identify the different Nasa Yuwe linguistic units, and to avoid problems with characters not included on the Castilian keyboard, it was decided to graphically represent each one of the 69 graphemes in the interface.

The first keyboard layout was based on a linguistic representation that was presented in a workshop about Nasa Yuwe, held in March 2003 at the *resguardo* of Pueblo Nuevo and led by Tulio Rojas Curieux and Abelardo Ramos.

Not all the children speak Nasa Yuwe, so the game must allow users to change the interface into Castilian, even if they will continue to guess words in Nasa Yuwe.

Nasalization is represented by a tilde (~) over the nasal vowels: *ĩ*, *ẽ*, *ã*, and *ũ*. There are some technical limitations in encoding character sets. *Latin-1*, commonly used in computer systems, only includes *ã* (ISO/IEC, 1998). *Unicode*, at least version 1.1, must be used to have full support (The Unicode Consortium, 1995).

6.3.1.3 Environment and Technology
The Nasa environment is mainly agricultural and has a strong link with Nature. Instead of the traditional *Hangman* as stimulus in the game, an important element in the Nasa culture and staple in the Nasa diet was proposed: *maize*. When the player suggests a letter that occurs in the word, it will appear in the corresponding spaces. If the player guesses wrong, the maize plant will loose some part such as a leaf or some seeds. The game continues in this way until the player finds the whole word or the plant looses all its parts.

6.3.1.4 Temporal Notions Time is less important in the Nasa culture than to Western cultures. In a Western context, a limitation in the playing time is acceptable, but it is not for the Nasa. A Nasa player would not mind spending a considerable amount of time to find a word.

A static interface (see Figure 6.6) was designed in this cycle. This was the first version of the prototype, which was evaluated through interviews with Abelardo Ramos and Tulio Rojas Curieux.

Abelardo's first reaction when he saw the prototype was: "What pretty maize!" referring to the picture used in the interface. Developers interviewed both evaluators and found some changes needed for the next version; for example, the keyboard distribution was not very appropriate. The empty spaces between buttons did not have any meaning. In general, a new layout for the interface was needed.

To determine the number of chances that the game should give the players to find a whole word, Abelardo played *Hangman* in Nasa Yuwe on paper. The result was an average of 16 guesses, which represents a medium level of difficulty; not too easy and not too difficult.

6.3.2 Cycle Number 1

The interface was modified according to the results of the previous cycle. The outcome was the functional prototype shown in Figure 6.7, which was tested by six Nasa girls, between 14 and 17 years old. This prototype was evaluated through *card sorting* and *thinking aloud* tests.

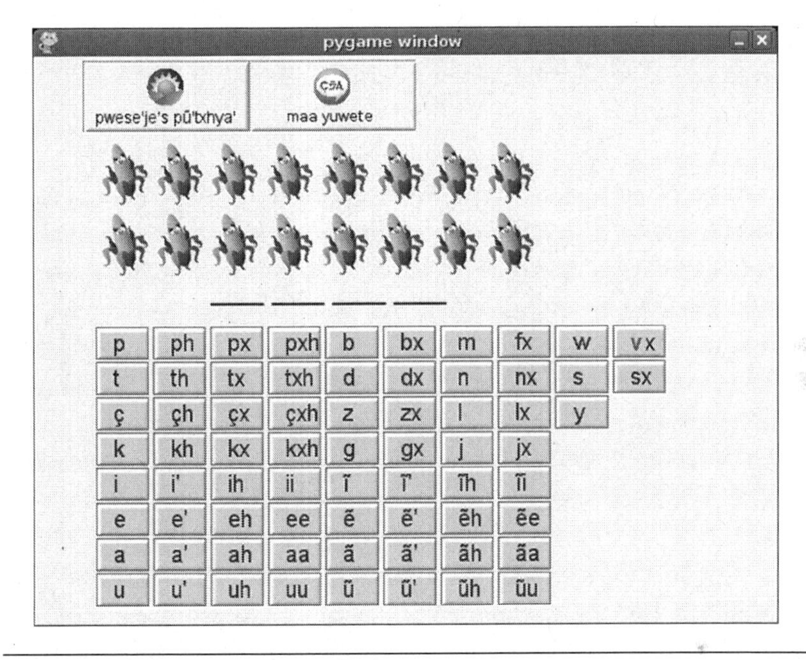

Figure 6.7 *Cut pwese'je* prototype No. 1.

In this new cycle, some additional cultural characteristics were found and taken into account:

6.3.2.1 Environment and Technology With respect to the game stimulus, the initial idea was that when the player proposed a "wrong" letter, the maize plant would lose a leaf or some seeds. However, since we determined that the number of chances for guessing the word was 16, we had to look for another option, because representing that number of transformations in a single plant was difficult, and maybe not very visible for the player. So instead, we chose to use 16 plants, one of which disappears each time the user "fails." We arranged them in two rows of eight plants each, as shown in Figure 6.7.

The rural environment is also present in the icon design. There were two options for the "*Start a new game*" button: a black paper (as is used in some computer games and in the first prototype) and a sunshine landscape icon. The Nasa girls chose the latter, shown in Figure 6.8, following their

Figure 6.8 *Pwese'je's takh"* (start a new game) icon.

preferences regarding weather: "It is when the sun is shining that we want to go out to play, not when it is raining."

6.3.2.2 Nonverbal Signs A change from the last version was that the maize pictures used were humanized, with eyes and a nose to make them more attractive to children. But this idea, which might be acceptable in a western context, was not well received by Nasa people. They prefer images and ideas that are as natural as possible, like the picture used in the first prototype.

6.3.2.3 Language Comments from the initial evaluation indicated that it was not easy to find certain letters in the graphical keyboard. It was too cramped. A better distribution was needed.

6.3.2.4 Social Organization It is typical for software interfaces to provide information *about* themselves, giving credit to people involved in the development process. In our first prototype, we also included a *Help* menu where *About* information was included (see the *pu'çxhi* menu in Figure 6.6). But this sense of credit does not exist in the Nasa culture, to the point that there is not even a word in their language with that meaning.

Note the lack of individual credit in references (Cátedra Nasa Unesco, 2001; CRIC, 1996a; CRIC, 1996b; CRIC and Rappaport, 2004) Nasa people worked on them, but they refer only to the indigenous organizations who represent the community.

To be consistent with the Nasa collective sense, the credit information was removed from this prototype, and with it, the help menu.

The Nasa social organization, and more precisely collectivism, was also present when evaluating the prototype. Individual evaluation activities like *card sorting* or *thinking aloud*, used in this cycle, were not so fruitful. Collective tests were more natural and better adapted to participation.

The game was in general well received by the girls who played with it, but some improvements were needed. Summarizing, the main changes for the next version were:

- Improvement of the keyboard distribution, and
- Removal of the human characteristics from the maize plants.

6.3.3 Cycle Number 2

Besides the changes proposed in the previous cycle, additional interviews with Abelardo Ramos and Tulio Rojas resulted in significant insights which allowed us to propose the interface in Figure 6.11. This prototype was evaluated in a collective meeting with 30 teachers from Nasa schools.

Cultural aspects, learned in the interviews and final evaluation, are summarized here:

6.3.3.1 Nonverbal Signs At this point, the developers learned about three important nonverbal signs in the Nasa culture: the *ũhza yafx*, the spiral, and the number system.

Ũhza yafx: This is a rhombus-shaped symbol, shown in Figure 6.9, which is well known in the Nasa culture as an essential part of the *cosmovisión*. In the world, there are four spirits of the cosmos, represented by four *truenos* (thunderbolts). The first one is found at the top, another one is on the bottom, and the other two are on the left and right, corresponding to the rhombus' corners. The Nasa world is divided in two, but they are just one unit; all this is present in the *ũhza yafx*.

Figure 6.9 *Ũhza yafx.*

To our knowledge, bibliographical references do not exist for this symbol. All information about it was learned through interviews with experts on the Nasa culture.

Ũhza yafx was used as a layout for the maize plants. It was easily identified by Nasa teachers when testing the game and was a more satisfying option than the two rows.

The *spiral* (see Figure 6.10) is another common symbol, which represents time and life development, and it is also used as an analogy to the different social interaction levels of hearth, house garden, *resguardo*, etc. (Rappaport, 2005). This was another symbol which could be used in the interface.

As a player proposes a letter that does not occur in the word, he or she loses a maize plant. When designing the interface, we wondered about the order that we should follow to remove the plants from the interface. Should it start at the center and end with the four

Figure 6.10 The spiral, Nasa common symbol.

corners since the four main gods are there? Or should it start on the borders, or just use a random path? We presented a prototype in the evaluation meeting with the first. When the users were asked to provide their opinion, they answered "Only a dog starts harvesting from the center; a good Nasa starts from the borders and ends in the center." They accepted the spiral as a good way for eliminating maize plants.

The Nasa number system. Abelardo Ramos offers a description about Nasa numerals (Ramos Pacho, n.d.):

- *Teeçx* (1). This is the number *dxi'pji'pmesa* "without a couple," such as the sun or the moon.
- *Je'z* (2). Has a binary link, *pdxi'p* "pair."
- *Tekh* (3). "A pair and another unit element." Connected with the three stones used in the hearth of the house.
- *Pahz* (4). Derived from *je'z pdxi'p* "two pairs." Related to the *ksxa'w* "Spirits of the Cosmos," who live in the space represented by the rhombus figure *ũhza yafx*.
- *Tahc* (5). Derived from "two pairs and a unit."
- *Setx* (6). Derived from "three pairs."
- *Sa't* (7). Derived from *tekh pdxi'psaki' teeçxidxi'pmesa* "three pairs and one *jigra**."

* Jigra: Nasa typical handcrafted bag.

The number 4 has a link with the cosmos and importance in the culture. There is also a sequence that comes from the *ũhza yafx*: *4-2-4-1-4*, which is used in alphabet instruction: p, t, ç, k-m, n-b, d, z, and g described below.

We took these numerical concepts to design the graphical keyboard. It includes four 4 × 4 blocks of letters. Additionally, we used some Nasa language and space concepts to further structure the elements in the interface.

6.3.3.2 Language At schools, the Nasa Yuwe alphabet is divided and its teaching follows a particular order: after vowels, children learn four basic consonants, namely *p, t, ç, k,* followed by two, namely *m, n*, then, in turn, *b, d, z, g,* and so on. The number of consonants presented in each group, *4-2-4*, corresponds to a sequence from the Nasa number system, which comes from one of the cosmos conceptions, represented in the *ũhza yafx*. We followed the alphabet learning sequence to organize the keyboard, so that vowels and the four basic consonants could be found easily.

6.3.3.3. Space This prototype was well accepted by the teachers in the testing meeting. During the test, they expressed that it was fun, and they easily identified Nasa symbols, such as the *ũhza yafx* and the number four in the keyboard design. Letters were accessed more easy than in the previous version. However, yet further improvements are being considered in the current development cycle. (See Figure 6.11.)

6.3.4 Current Cycle (Cycle 3)

Following are some suggestions gained in the previous evaluation on how to improve the interface and more closely match the Nasa cultural expectations.

- Nasa people prefer "natural contexts." The maize graphic used in cycle 1 was humanized. It had eyes and a nose, which had to be removed in order to make the maize as natural as possible.
- In the meetings where we presented the game to the groups of users, we used descriptions such as "You have to save the plants," or "If you do not find the word, the maize will disappear." We perceived that the users did not feel entirely

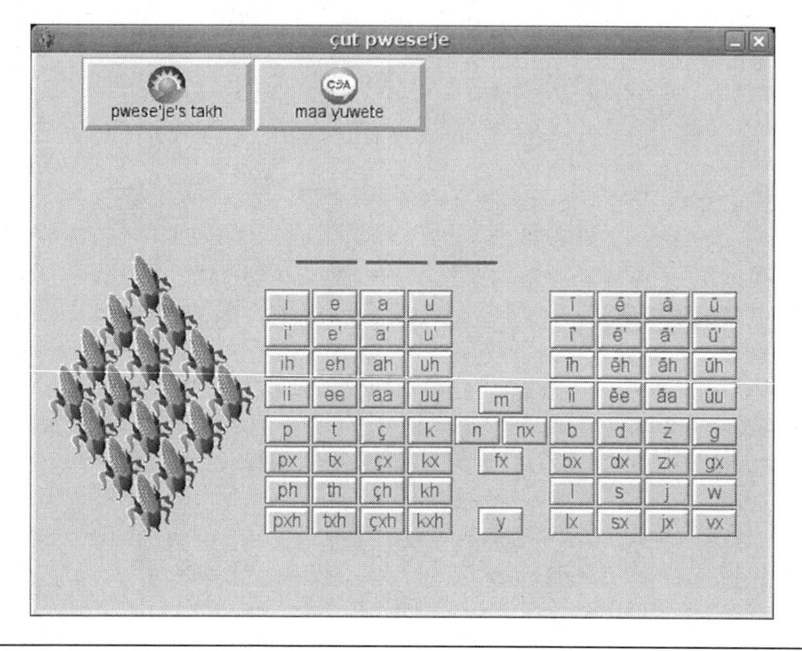

Figure 6.11 Current version of the *çut pwese'je* game.

comfortable with these ideas. We received a suggestion from a University of Cauca professor to include a squirrel that would steal the maize plants when the user inputs an "incorrect" letter; this is a more natural action and therefore, more likely to be acceptable to the Nasa people.

- In the future, we may take into account space structuring and energy movements. If the stealing squirrel is a "negative" element, it should leave the interface on the right, as disharmonies do. As a "positive" action, the player should drag the letters to the left side of the game.

6.4 Conclusions

The six dimensions—*Language*, *Space*, *Environment and Technology*, *Social Organization*, *Temporal Notion*, and *Nonverbal Signs*, are the result of comparing existing cultural models with knowledge about and experiences with the Nasa Colombian culture. These dimensions cover a wide range of cultural variables, which should be considered in the design of interfaces for a given culture, and user-centered in computer tool software development processes. Such information can help software developers to understand culture and how it impacts people's expectations surrounding interactions whether with other people or with machines.

We have described how the Nasa culture can be characterized in this model, and how their cultural variables impact both the design and development process for "the maize game." Cultural differences and singularities may affect not only the user interfaces, but the whole development process. For example, we found that evaluation of prototypes for the Nasa people had to involve the community, not just individuals, because of the collective nature of their culture. The evaluations were influenced by the people's collective sense, as well as elders' and leaders' opinions and political positions.

Çut pwe'seje was an initial successful step towards producing suitable pedagogical software for the Nasa people. In the future, we will use the model to design a larger Nasa collective learning tool which will be tested in their classrooms.

In our context, we consider *etiquette* to reflect culture and its implications for interaction. We suspect that attention to culture and *etiquette* in design of human–computer interaction will increase the likelihood of producing appropriate interface interactions, symbols, and layouts that will be interpreted correctly, and be accepted.

During the second development cycle of the game, in order to present a more "friendly" interface, developers requested help from graphic designers who had little knowledge of the Nasa culture. The graphic designers produced several proposals. The pictographs used "prehistoric" pictographs usually found on stones (shown in Figure 6.12). However, they were not Nasa. When designers showed these proposals to Abelardo and Tulio, they confirmed that "This is not Nasa. If you show this to Nasa people, they will reject the game; they will even not play."

We hope that this work will help others efforts in making computing accessible to a wider range of cultures, including minority and indigenous people who might otherwise find most computing applications "inaccessible" even when computers are available.

Figure 6.12 Non-Nasa pictogram.

Acknowledgments

Special acknowledgments for Angela Consuelo Checa, who was one of the game developers and part of the process that triggered the idea for the game.

Help from Abelardo Ramos has been essential. Without him, this project would not have been possible. His contributions as linguist, Nasa representative, and primary contact with the Nasa community have been invaluable. We would also like to offer a special thanks to Aimée Johansen for improving the nonnative English used in this chapter.

References

Cátedra Nasa Unesco. *Cómo vivían nuestros mayores. Yatchë'wesxa kapiya'hatx u'hunxisa: Los mayores nos enseñan a través de sus historias vividas.* Asociación de Cabildos Indígenas del Norte del Cauca—ACIN, 2001.

Cátedra Nasa Unesco. *Las Luchas de los mayores son nuestra esperanza.* Asociación de Cabildos Indígenas del Norte del Cauca—ACIN, 2001.

Checa Hurtado, Á. and Ruano Rincón, S. *Lineamientos para la adecuación de IGUs en el ámbito de la cultura indígena Paez.* Universidad del Cauca, 2006. Trabajo de grado.

Consejo Regional Indígena del Cauca—CRIC. Nasa yuwe' walasa'—nuestra lengua es importante, 1996a. Video.

Consejo Regional Indígena del Cauca—CRIC. Nasa tul—la huerta nasa, 1996b. Video.

Consejo Regional Indígena del Cauca—CRIC and Rappaport, J. *Qué pasaría si la escuela.* PEBI-CRIC, 2004.

de Souza, C.S. *The Semiotic Engineering of Human-Computer Interaction.* MIT Press, Cambridge, MA, 2005.
Hall, E. *The Dance of Life: The Other Dimension of Time.* Anchor Press, New York, 1989.
Hall, E. *Understanding Cultural Differences: Germans, French, and Americans.* Intercultural Press, Yarmouth, ME, 1990.
Harrison, R. *Beyond Words: An Introduction to Nonverbal Communication.* Prentice-Hall, Englewood Cliffs, NJ, 1974.
Hayes, C., Miller, C., and Pande, A. Etiquette in human computer interactions: What does it mean for a computer to be polite? or who needs polite computers anyway? In *AAAI Fall Symposium*, 2002.
Hofstede, G. and Hofstede, G.J. *Cultures and Organizations: Software of the Mind.* McGraw-Hill, New York, 2005.
Hoft, N. Developing a cultural model. In *International User Interfaces.* Wiley Computer Publishing, 1996.
ISO/IEC. *8859-1:1998 Information Technology—8-bit Single-Byte Coded Graphic Character Sets—Part 1: Latin Alphabet No. 1*, 1998.
Kluckhohn, F. and Strodtbeck, F. *Variations in Value Orientations.* Row, Peterson, 1961.
Lee, K.-P. *Culture and Its Effects on Human Interaction with Design. With the Emphasis on Cross-Cultural Perspectives between Korea and Japan.* PhD thesis, University of Tsukuba, 2001.
Portela, G.H. Cuerpo y cosmos en el rito paez. 2008.
Ramos Pacho, A. Numeración y neonumeración en nasa yuwe. *Cartilla Semillas y mensajes de etnoeducación çxayu'çe*, (9), n.d.
Rappaport, J. *Intercultural Utopias: Public Intellectuals, Cultural Experimentation, and Ethnic Pluralism in Colombia.* Latin America Otherwise. Duke University Press, Durham, NC, 2005.
Rojas Curieux, T. Desde arriba y por abajo construyendo el alfabeto nasa. la experiencia de la unificación del alfabeto de la lengua páez (nasa yuwe) en el departamento del cauca—Colombia, 2002. http://lanic.utexas.edu/project/etext/llilas/cilla/rojas.html.
The Unicode Consortium. Unicode 1.1 composite name list, 1995. http://www.unicode.org/Public/1.1-Update/UnicodeData-1.1.5.txt.
Trompenaars, F. *Riding the Waves of Culture: Understanding Cultural Diversity in Business.* Nicholas Brealey, London, 1993.
Valencia, J.H.G. Lugares y sentidos de la memoria indígena paez. *Convergencia*, (21):167–202, 2000.
Victor, D. *International Business Communication.* New York: HarperCollins, 1992.
Wikipedia. Gran minga por la vida, 2009. http://es.wikipedia.org/wiki/Minga Indígena—Accessed on June 24, 2009.

PART III

Etiquette and Development of Trust

7

Network Operations

Developing Trust in Human and Computer Agents

MARK T. DZINDOLET, HALL P. BECK, AND LINDA G. PIERCE

Contents

7.1 The Human in Network Operations

We know a great deal about when humans will rely on other humans, and how trust develops and is maintained between them. We have a theory describing human processes and biases that can help explain how people learn to trust others, and when and why they will overly trust or distrust them (Holmes 1991; Larzelere and Huston 1980; Rempel, Holmes, and Zanna 1985). Do these same principles apply to human reliance on automation or is trust between humans and automation fundamentally different than that which develops between humans (Lee and See 2004; Madhavan and Wiegmann 2007)? What are the factors that encourage appropriate automation reliance and how can these factors be used to influence system design? One of the factors that may affect trust in both human–human and human–automated interactions is etiquette (Parasuraman and Miller 2004). The importance of human–machine etiquette (HME) in understanding automation reliance is only recently being realized (Miller 2002).

7.2 The Military and Network Centric Operations

The transition of military technologies and operations from an industrial age to an information age is fueling much of the military research in networks and automation reliance. To better capitalize on information age technologies, the U.S. military is implementing Network Centric Operations (NCO). NCO is a networking strategy to improve the efficiency and effectiveness of operations by improving the speed of communications and the likelihood of shared situation awareness (Wilson 2007). We contend that information sharing, situation awareness, and self-synchronization, all tenets of NCO, will not be insured by merely having faster access to more information, but rather through a better understanding of how and when humans will use and interact with information technology to improve collaboration and decision making.

A key enabler of NCO for the U.S. Army is the Future Combat Systems (FCS), a conceptual, interdependent, networked system of systems. The FCS will support military functions across the spectrum of move, shoot, and communicate. When fielded, FCS will represent an alliance of humans and machines previously not seen or possible in military operations (Lickteig, Sanders, Durlach, Lussier, Carnahan, and Sauer 2003). In fact, if successful, it is conceivable that FCS may blur the lines between humans and automation. Ideally, a person would have difficulty answering the question, "Was this information fused by a human, an intelligent agent 'bot,' or both? Was that alert from an automated system, a team member, or my commander?" Essentially, automated aids will extend the team by operating as automated team members. How will the social processes and biases that influence human–human interactions extend to interactions between humans and machines? We believe that understanding human–machine etiquette and its role in engineering appropriate levels of trust and cooperation between humans and computer assistants may be important in creating high functioning teams of the future.

The idea that military operations will be improved by networking information, humans, and weapon systems is not new. However, much of the investment in NCO and FCS has been in information technology and the development of technical interoperability rather than in the sociotechnical integration that will be necessary to realize the true power of networks (Alberts and Hayes 2006; National Research Council 2005; Shattuck, Miller, and Miller 2006). Technical interoperability will be a necessary, but not a sufficient, requirement for networking. Additionally, Alberts and Hayes (2006) detail a need for semantic interoperability, defined as the capacity to fully comprehend one another, and "cooperability" or the willingness to interact and the desire to communicate clearly. In this work, we extend the notions of semantic interoperability and "cooperability" beyond human–human interactions to human–machine interactions. Thus, the goal of this chapter is to provide a better understanding of "human" processes and biases in the use of technology, and to propose system guidelines that will facilitate the design of information technology which can realize the potential of networked operations. The objective is to create a true alliance between humans and automation that enables, rather than impedes, collaboration and decision making in the FCS battle command (Lickteig, Sanders, Durlach, Lussier, Carnahan, and Sauer 2003).

To better understand both human–human and human–machine relationships, and develop design guidelines for task allocation, Lickteig, Sanders, Durlach, Lussier, Carnahan, and Sauer (2003) conducted a series of experiments using prototype FCS and operational concepts in a simulated but realistic setting. They worked closely with potential FCS users and employed both objective and subjective means to document functions and tasks required by an FCS command group. They assessed perceived effectiveness of each of the proposed automated systems, calculated the number of interactions between humans and systems by function and command group position, and evaluated workload and training. They showed that some features of the FCS command and control system were used more frequently by individual command group members and across functions. They also found that the perceived effectiveness of the features varied, the workload was high, and training was incomplete. This work was preliminary but important because it established a baseline for further user-centered design and development of the human–automation interface, which is so critical to FCS operations.

7.3 Defining Key Terms

Miller (2002) defined *etiquette* in this way:

> By "etiquette," we mean the defined roles and acceptable behaviors and interaction moves of each participant in a common "social" setting—that is, one that involves more than one intelligent agent. Etiquette rules create an informal contract between participants in a social interaction, allowing expectations to be formed and used about the behavior of other parties, and defining what counts as good behavior (p. 2).

Understanding the "informal contract" that human operators have with their automated decision aids should help researchers to predict when operators will appropriately rely on their automated aids.

An assumption underlying work in automated decision aids is that the human–computer dyads will perform better than the humans or the computer aids would have performed alone. Although Dalal and Kasper (1994) have found support for this underlying assumption, other researchers have found instances in which human operators do not appropriately rely on automation. Parasuraman and Riley (1997) identified misuse and disuse as two forms of inappropriate reliance. *Disuse* occurs when human operators underutilize their automated aids. Parasuraman and Riley (1997) describe many real-world incidences in which disastrous results occurred because people ignored automated warning signals. *Misuse* occurs when human operators overly rely on their automated aids. Relying on automation when manual performance would have been better is an example of misuse. We believe an understanding of human–computer etiquette and its impact on trust and cooperation can help researchers predict when human operators will be likely to disuse, misuse, and appropriately use automation. This information can be used by software designers and trainers to create optimally functioning human–automation teams. We agree with Miller that "… by placing the system to be designed in the role of a well-behaved human collaborator, we gain insights into how users might like or expect a system to act and how its actions might be interpreted *as a social being* that rarely come from any other source (Miller 2002, p. 3)." Miller (2002) identified some of the characteristics of etiquette which may be useful to understanding human–computer interactions. We will use some of these characteristics to organize this chapter. Specifically, we will review the literature in general and some of our own experimental data in particular to test some of Miller's human–computer etiquette claims and provide some suggestions for future work.

7.4 Evaluation of HME Claim: Etiquette Is Implicit, Complex, and Implies a Role

Miller (2002) suggests that etiquette is implicit, complex, implies a role, and even conveys intention. The processes through which humans form impressions and expectations of others, form hypothesis as to why others act the way they do, and make decisions about how to respond are not obvious to the casual observer. We will begin by exploring how these processes influence human–human relationships, and then explore the extent to which similar processes occur in human-automated "relationships." The expectations that humans have for other humans are likely to drive some of the expectations humans have for machines (Miller, 2004). This is likely to affect human–machine interactions and have implications for how HME may affect automation reliance.

7.4.1 Formation of Impressions and Expectations of Human Partners

7.4.1.1 Impressions Based on Physical Appearance In human–human relationships, impressions are based, in part, on appearances (see Klein's chapter). People make trait inferences from facial characteristics, and they do this very quickly. Willis and Todorov (2006) found participants' ratings of a stranger's trustworthiness, competence, likeability, aggressiveness, and attractiveness based only on viewing a picture of the person's face for 100 milleseconds were highly correlated with ratings made without any time constraints. In other words, in one tenth of a second (or less), participants in the study formed expectations about the core traits of a stranger. Surprisingly, the strongest correlation Willis and Todorov (2006) found was *not* for attractiveness (a trait that is truly dependent upon the facial features of the stranger), but for trustworthiness! Willis and Todorov hypothesize that, from an evolutionary perspective, accurate trustworthiness judgments are necessary for human survival, and suggest that neuroimaging studies have found trustworthiness judgments are related to activity in the amygdale (Winston, Strange, O'Doherty, and Dolan 2002). Similarly, Bar, Neta, and Linz (2006) found reliable threat judgments from exposure to neutral faces for only 39 ms. And Berry (1990) found that such judgments may not only be reliable, they may be accurate. However, these impressions may be affected by communication medium. Specifically, Fullwood (2007) found participants made less favorable impressions (concerning likeability and intelligence) of partners with whom they communicated via video conferencing than of partners with whom they communicated face-to-face.

Does the visual appearance of an automated aid (e.g., the interface) similarly affect a human operator's ratings of the machine's trustworthiness or competence? Can behaviors of the automated aid such as use of appropriate etiquette additionally affect these ratings? The answer to these questions may have design implications; if we can better understand how to design automated aids that are deemed trustworthy people may be more willing to accept, use, and rely on them.

7.4.1.2 Impressions Based on Rumor and Other Prior Knowledge People also form impressions of others based on very limited, even unreliable, information, such as from rumor. In a classic social psychological study, students were provided with information about a substitute instructor. Specifically they were told the following:

> Mr. ____ is a graduate student in the Department of Economics and Social Science here at M.I.T. He has had three semesters of teaching experience in psychology at another college. This is his first semester teaching Economics 70. He is 26 years old, a veteran, and married. People who know him consider him to be a rather **** person, industrious, critical, practical, and determined (Kelley 1950, p. 432).

The stars in the above scenario were replaced with the word "cold" for half the participants and "warm" for the others. At the end of the class period, students were asked to rate the instructor on several dimensions. Students led to expect that the temporary instructor would be warm rated him as more informed, sociable, popular, humorous, considerate of others, and better natured than those led to expect the instructor would be cold. In addition, students expecting the instructor to be warm were significantly more likely to interact with the instructor than those expecting him to be cold. Therefore, humans' impressions of others tend to be formed very quickly and on limited information. If humans' impressions of automated systems are formed with equal rapidity and limited information then initial etiquette exhibited by a computer system may be especially important.

7.4.1.3 Impression Persistence Initial impressions in human relationships are long lasting (primacy effect; Asch 1946). Several hypotheses have been created to explain why first impressions are powerful and long lasting. One explanation is that people are more attentive to initial information about a person than they are to later information. This may be due to the fact that people's very survival may depend on their accuracy and speed in determining if another will help or harm them. Therefore, people extend much cognitive energy initially in the interaction trying to classify and categorize their partner as harmful or safe, and to reduce the interaction's uncertainty and unpredictability. Once the partner is categorized, uncertainty declines. At this point, the cognitive energy devoted to interacting with the partner can be reduced and distributed elsewhere. For this reason, people may be less attentive to later information relative to initial information. Therefore, the initial information is more powerful in impression formation than later information.

In its extreme form, people cling to their initial beliefs even in the face of contradictory evidence. This has been dubbed the *belief persistence effect* (Downey and Christensen 2006; Ross, Lepper, and Hubbard 1975). Downey and Christensen (2006) provided participants with a photograph and some background information of the person photographed. After viewing the photograph and reading the information, participants were to rate the person's attractiveness, rate how much the participant liked the person in the photo, the possibility of developing a friendship with the photographed person, and the person's perceived truthfulness. The background information was manipulated such that the person in the photograph was depicted as morally neutral (e.g., a member of a band, currently attending college) or immoral (e.g., history of theft, incarceration). After reading this information but before making the ratings, half of the participants in each condition were told that the background information with which they had been provided was incorrect. These participants were provided with new and correct background information such that the neutral information was replaced with the immoral information and the immoral information was replaced with the neutral information. Participants were provided with additional time to read the new information. Other participants were not told that their initial information was incorrect.

Logically, the ratings of the photographed person should not differ between participants provided with neutral information—whether this replaced initially incorrect "immoral" information or not. However, participants who initially read the immoral background rated the person more negatively than those who had just been provided with the neutral information. Participants clung to their initial impressions even when they were told that the initial information on which the impression was formed was incorrect.

7.4.1.4 Positive and Negative Biases Downey and Christensen (2006) also found evidence for a negativity bias in their data (e.g., a tendency to attribute greater weight to negative than positive behaviors and character traits; Fiske [1980]). The lowest ratings were given by participants who were provided with "immoral" background information after initially reading the neutral information. Therefore, the corrected negative information had a stronger impact on the final ratings than corrected neutral information.

Skowronski and Carlston (1987) hypothesize that negative behaviors are given more weight in impression formation only when negative behaviors are more diagnostic than positive behaviors. For example, when assessing morality, negative behaviors are more diagnostic. Immoral people engage in many positive behaviors due to conformity pressure, but moral people perform few negative behaviors. Consistent with this hypothesis, they found a negativity bias in impression formation concerning morality assessments. Additionally, they suggest a positivity bias, or a tendency for people to attribute greater weight to positive than negative behaviors and traits, should occur in situations in which positive behaviors are more diagnostic than negative behaviors. For example, they suggest that when assessing ability, a positivity bias should occur. High ability people can perform poorly due to fatigue or lack of motivation; however, low ability people cannot perform well. Therefore, a high performance is diagnostic of high ability more than a low performance is diagnostic of low ability. Consistent with this hypothesis, Skowronski and Carlston (1987) found a positivity bias in impression formation concerning ability assessments. Finally, Skowronski and Carlston (1987) suggest that extreme behaviors are more diagnostic than moderate behaviors, and therefore, will play a large role in impression formation.

In summary, people make impressions of other people very quickly (even as little as 39 ms), often based on very limited information (e.g., a picture, a short description). The first impression is long lasting and persists in the face of contradictory evidence—even when the information used to form the initial information is found to be false. In addition, impressions seem to be driven by diagnostic information leading to an extremity bias for all dimensions, a positivity bias on ability dimensions, and a negativity bias on morality dimensions.

7.4.2 Impression Formation and Expectations of Automated Partners

Do people form impressions of computers according to patterns similar to the way they form impressions of other humans? Do they form impressions of automated aids rapidly, based on limited information, such as a short description? Does the appearance or "look" of the decision aid play a role in impression formation? Do tendencies in judging ability or morality even apply to computers?

Adding a human-looking element to a computer aid (e.g., an animated pedagogical agent, see Bickmore chapter) may increase interest in the task and (sometimes) has been found to increase learning (Atkinson 2002; Moreno, Mayer, Spires, and Lester 2001). Even computers that do not "look" human are treated as social agents (Nass, Moon, Fogg, and Reeves 1995; Reeves and Nass 1996). "Nonpersonified" automated aids, without facial expressions, speech, or gestures, can affect human operators' perceptions and appropriate use of the aid (Parasuraman and Miller 2004). Dzindolet, Peterson, Pomranky, Pierce, and Beck (2003) set out to determine the initial expectations people had of an automated decision aid. Their experiments are described below.

7.4.2.1 Experiment 1: Initial Impressions of an Automated Aid

Purpose: Dzindolet et al. (2003) performed a study to determine the expectations human operators had regarding a hypothetical automated decision aid about which they had extremely limited information.

Method: Fifteen Cameron University students were told they would view 200 slides displaying photographs of Fort Sill terrain on a computer screen. In about half of the photographs, a soldier, in varying degrees of camouflage, would be present (Figure 7.1 presents a sample slide). After each slide, subjects were asked to indicate whether or not they believed the soldier was present in the photograph, and the confidence they had in their decision. Next, they would view the decision reached by an automated decision aid named "Contrast Detector" which was designed to automatically use shading contrasts to determine the likelihood of a soldier's presence.

All participants were told that their automated aid, "Contrast Detector," was not perfect. Participants performed four very easy practice trials for which the aid supplied the correct decisions and were given the opportunity to ask questions. After performing the

Figure 7.1 Sample soldier slide.

four practice trials, participants were asked to estimate their automated aid's expected performance on the upcoming 200 trials.

Results: With limited information about an automated decision aid, participants easily formed an impression of the automation. Consistent with the positivity bias that, when assessing ability, positive behaviors are more diagnostic, participants expected the automated aid to perform very well (M = 7.18 on a 9-point scale). Although individual differences exist (Lee and Moray 1992; 1994; Singh, Molloy and Parasuraman 1997), people form impressions of automated aids based on limited information. How similar are the actual impressions formed of automated aids to those formed of humans?

7.4.2.2 Experiment 2: Comparison of Initial Expectations of Human and Computer Partners

Purpose: The purpose of this study was to compare the initial expectations human operators had concerning an automated "partner" with the expectations the human operators held toward a human "partner."

Method: Half of the 68 Cameron University students in this study were treated exactly like those in the Dzindolet et al. (2003) study described above. However, the other participants were led to believe that the decisions they would view were made by the prior participant. In other words, some participants believed they were paired with an automated partner; others believed they were paired with a human partner. After four easy practice trials, participants estimated their and their partner's (either human or automated) expected performance in the 200 upcoming trials.

Results: Participants rated the expected performance of the contrast detector more favorably than that of the prior participant. Those expecting to work with the contrast detector estimated it would make, on average, less than half as many errors as those expecting to work with the prior participant estimated the human partner would make. Although people form impressions of both human and automated aids quickly and based on limited information, the actual expectations they form appear to differ. Given identical information, people expected automated partners to outperform human partners.

This result was not due to participants' belief that their human partners' performance would be inferior to their own. Participants expecting to work with the contrast detector estimated that the automated aid would make far fewer errors than they would make themselves; participants expecting to work with the prior participant estimated the prior participant would make about the same number of errors as they. Therefore, not only did the participants expect their automated partners to outperform human partners, they expected their automated partners to outperform themselves! A conclusion drawn from this study was that people hold a bias in favor of automation; the participants predicted the automated aids would perform better than the human aids. Similarly, Dijkstra, Liebrand, and Timminga (1998) found that students judged advice from an automated expert system to be more rational and objective than the same advice from a human advisor.

Summary: Extending our understanding of etiquette in human relationships to human–computer relationships has proved useful. The processes by which humans form impressions of other humans and of automated aids seem to be similar. People quickly form impressions of human and automated aids based on limited data. However, their impressions of human and computers are not identical; people appear to be influenced by beliefs they hold about differences in the abilities of humans and computers. Additionally, this approach has revealed many questions that remain unanswered: Do human operators suffer from the belief persistence effect in their impressions of automated aids? What are the features of automation that humans deem diagnostic and, therefore, are likely to lead to biases in impression formation? What are the boundary conditions of the bias toward automation and why does it exist? How might human–computer etiquette (e.g., interface appearance and system behavior) affect initial impressions of trustworthiness and ability, belief persistence, and the bias toward automation?

7.4.3 Causal Attributions with Human Partners

People use their impressions of other people to try to understand and explain why others act they way they do. How people explain others' behaviors has been extensively studied by social psychologists. Attribution theorists (e.g., Heider 1958; Jones and Davis 1965; Kelley 1972) hypothesize that people categorize the cause of behavior as either internal or external to the person performing the behavior. If the behavior is thought to be the result of a person's thoughts, intentions, abilities, or personality, an internal attribution is made. If the behavior is thought to be the result of situational forces, an external attribution is made. According to Kelley's Covariation Model (Kelley 1972; Kelley and Michela 1980), people make causal attributions by examining the likelihood of the behavior when the cause is present and absent. If the behavior and possible cause occur together more often than the behavior occurs in the absence of the possible cause, then the two can be said to covary, and people are likely to infer a causal relationship between them. People focus on three types of information to make an internal or external attribution: (1) distinctiveness, (2) consensus, and (3) consistency.

Distinctiveness is the extent to which it is unusual for a person to exhibit a given behavior across a variety of contexts. Highly distinctive behaviors tend to be attributed to external sources; low distinctiveness behaviors tend to be attributed to internal forces. *Consensus* is the extent to which other people are exhibiting the behavior in a given context. High consensus behaviors tend to be attributed to external forces; low consensus behaviors tend to be attributed to internal ones. *Consistency* is the extent to which a person repeats the same behavior in the same context, but at a later time. If consistency is low, an attribution is not typically made; the behavior is thought to be a coincidence or an aberration. However, if consistency is high, then an attribution will likely be made.

Using Kelley's Covariation Model, a person would probably make an accurate determination as to the cause of another's behavior. However, the determination would require the use of cognitive resources. To determine why a hypothetical partner, Karen, made a decision error while performing the soldier detection task described in the experiments above, one would need to search one's memory in order to assess how often Karen makes decision errors as she performs other tasks (distinctiveness), and how often other people make decision errors while performing the soldier detection task (consensus). Additionally, one would need to attend to whether Karen makes many errors while performing the soldier detection task in the future trials (consistency). Instead, people often make attribution errors because they are not willing or able to put in that cognitive effort.

People are biased toward making internal rather than external attributions; this is called the *fundamental attribution error* (Ross 1977) and it is hypothesized that this is because dispositional characteristics are oftentimes more salient than situational characteristics (Krull and Dill 1996; but see Robins, Spranca, and Mendelsohn 1996). When explaining others' behavior, people are likely to underestimate the effect of the situation and overestimate the effect of the person's intentions, attitudes, abilities, or personality. However, when people are trying to understand their own behaviors, the bias is reversed; external attributions are more likely. To explain their own behavior, people are likely to overestimate the effect of the situation and underestimate the effect of personal attributes. This bias is called the *actor–observer effect* (Jones and Nisbett 1972; Karasawa 1995; Krueger, Ham, and Linford 1996).

In addition to the fundamental attribution error and the actor–observer effect, people make self-serving biases in their attributions. When faced with positive outcomes, people are more likely to make internal attributions; when faced with negative outcomes, people are more likely to make external attributions (Mezulis, Abramson, Hyde, and Hankin [2004]). Another self-serving bias is the *false consensus effect*, the belief that other people share our own attitudes more than they actually do (Ross, Greene and House 1977). In addition, people show an *optimistic bias* (Helweg-Larsen and Shepperd 2001); they overestimate their chances of experiencing common positive events (e.g., getting married and having children) but underestimate their chances of experiencing rare negative events (e.g., getting cancer). Humans exaggerate their contribution to a group product (appropriation of ideas, Wicklund 1989), overestimate the number of tasks they can complete in a given period of time (planning fallacy, Buehler, Griffin, and Ross 1994), and are overconfident in negotiations (Neale and Bazerman 1985). Finally, Hansen (1980) suggests that rather than seeking out distinctiveness, consensus, and consistency information, people will often reserve cognitive energy by generating a naïve causal hypothesis. They will create a possible explanation for why another behaved the way he or she did, seek out confirmatory evidence for their hypothesis, and, if such evidence is found, conclude that their hypothesis is correct. Although this effort-saving strategy is likely to lead to false conclusions, Hansen suggests people use this strategy more often than Kelley's more effortful strategy.

In summary, people can use distinctiveness, consensus, and consistency information to determine if another's behavior should be attributed to the person or the situation. However, to avoid the extensive cognitive effort that such a judgment would require, people often just assume that another's intentions correspond to behavior, or use naïve causal hypothesis testing. In addition, humans are plagued by many self-serving biases allowing them to attribute positive outcomes to their own personal properties and actions and negative ones to outside circumstances.

7.4.4 Causal Attributions with Automated Partners

Do people make attributions (and attribution errors) when they try to explain an automated aid's "behavior?" To what extent do human operators seek out distinctiveness, consensus, and consistency information to explain their automated aid's decisions? The human factors literature has largely ignored these questions. These questions need to be examined. When an automated aid makes an error, does the human operator try to determine if this aid makes similar errors in other situations (distinctiveness), if other aids make this kind of error in this situation (consensus), and if this aid makes this error in this situation at a later time (consistency)? Under what conditions is the human operator likely to extend this much cognitive energy? Does HME affect these processes? By addressing these questions, human factors researchers will be better able to provide guidance to system designers.

When the automated aid makes an error, much like with human–human interaction, do human operators make an internal attribution and conclude that the aid is not reliable? Madhavan and Wiegmann (2007) indicate that human operators do, in fact, commit the fundamental attribution error. Researchers should examine the boundary conditions for this effect and determine the extent to which HME plays a role. To what extent do human operators fall prey to the other attribution biases when explaining automation? Madhavan and Wiegmann (2007) conclude that research has found that human operators will blame failures and errors on technology, but take credit with positive outcomes. Thus, they find evidence for the actor–observer effect. To what extent can HME affect the human operator's reactions? What could system designers do to the interface appearance or the system's behavior that would make these attribution errors more (or less) likely? These questions need to be addressed in the human factors literature.

How prevalent are self-serving biases in attributions with automated versus human aids? Do people expect their automated decision aids to share their opinions (false consensus effect) or is this reserved only for human "decision aids?" Do human operators extend the optimistic bias and the planning fallacy to their automated partners and overestimate the extent to which their human-automated team will succeed, at the same time underestimating the amount of time it will take to be successful? Or do they view the automated partner as a rival and compete with it? To what extent do people take credit for the performances of their automated aid (appropriation of ideas)? To what extent does HME affect these processes?

7.4.4.1 Experiment 3: Attributing Causality (Dzindolet et al. 2003)

Purpose: The purpose of the study was to determine the extent to which understanding *why* an automated decision aid might affect humans' trust in the automated aid and their willingness to rely on it.

Method: We used the soldier detection paradigm described earlier in this chapter. However, in this study, the 24 Cameron University students actually performed the task (rather than just providing estimates of expected future performance). To provide participants with experience with the automated aid (Contrast Detector), participants viewed 200 slides showing photographs of Fort Sill terrain, half of which contained a camouflaged soldier. Each slide was presented for only three fourths of a second. After viewing the slide, the participant indicated whether or not a soldier was presented in the slide. After making their own decision, the human operators were provided with the decision reached by the Contrast Detector. All participants were informed that the automated aid was not perfect. However, half the participants were provided with a hypothesis as to why their automated aid (the contrast detector) might make an error. Specifically, they were told, "The Contrast Detector will indicate the soldier is present if it detects forms that humans often take. Since nonhumans (e.g., shading from a tree) sometimes take human-like forms, mistakes can be made." Participants in the control condition were not provided with a hypothesis of why their automated decision aid might err.

After the 200 trials, participants were provided with cumulative feedback on the Contrast Detector's performance relative to their own. Half the participants were told that the automated aid made half as many errors as they had made; other participants were told that the automated aid made twice as many errors as they had made. After viewing this cumulative feedback, participants performed 100 key trials. On these trials, participants were provided with the automated aid's decisions *prior* to making their absent–present decision. Specifically, the human operators viewed the slide, then viewed the Contrast Detector's decision, and then made their decision. For these key 100 trials, it was possible for the human operators to rely on the decisions reached by the Contrast Detector. We examined automation reliance by comparing the p(error | aid correct) and the p(error | aid incorrect); see Dzindolet et al. Study 3 (2003) for more information about the analyses. After completing the trials, the participants completed a survey in which they were asked to rate the reliability and trustworthiness of the automated aid.

Results: Participants provided with a reason why the aid might make an error were more likely to rely on their automated aid. Unfortunately, this increased reliance occurred regardless of the reliability of the automated aid. Thus, providing an attribution for an aid's mistake led to inappropriate reliance on inferior automated aids. In addition, it was found that this relationship was mediated by trust, a factor that will be discussed later in this chapter. Providing a reason as to why the Contrast Detector may err led to increased trust in the Contrast Detector, and increased trust led to increased automation reliance.

7.4.4.2 Summary of Causal Attributions with Automated Partners Once again, extending our understanding of etiquette in human relationships to human–computer relationships has proven to be useful in understanding human–computer relationships and in generating research questions. Do human operators use completely different strategies for making causal attributions for automated aid's behaviors? Perhaps a new attribution theory will need to be created to understand attributions for automation. It is possible that such a theory may help our understanding of attributions for human behavior.

7.4.5 How Expectations Affect Interactions with Human Partners

Initial impressions, expectations, and attributions are important in understanding human relationships because they affect human interactions. First, people seek out information that is consistent with their expectations. This phenomenon has been dubbed *perceptual confirmation* (Copeland 1993). In addition, people actually create in their partners behaviors that are consistent with their expectations (*behavioral confirmation* or *self-fulfilling prophecy*, Copeland 1993). The self-fulfilling prophesy has been found in a variety of contexts among teachers and students, nursing home caretakers and their residents, and experimenters and animal subjects (Copeland 1993; Darley and Fazio 1980; Merton 1948; Rosenthal 1994). Harris, Lightner, and Manolis (1998) noted that much of the research that had found behavioral confirmation involved situations in which the person forming the impression of another (perceiver) had more power than the one who was being perceived (target). In an effort to determine if the self-fulfilling prophesy is due to the target–subordinate deferring to the expectations of the more powerful boss–perceiver (target deference hypothesis) or due to the boss–perceiver not paying attention to the target (cognitive economy hypothesis), they utilized a 2 (expectancy of target–subordinate ability: high or low creativity) X 2 (perceiver–boss' awareness of the status) X 2 (target–subordinate's awareness of the status) design. Participants performed a creative task (specifically, design an advertisement) in dyads; the entire interaction was videotaped and analyzed.

Harris, Lightner, and Manolis (1998) found evidence of perceptual confirmation. Regardless of the awareness conditions, perceivers who were led to believe their targets were highly creative rated the target's performance more positively than those led to believe their partner was less creative. The perceiver–bosses found evidence for their expectations even though this evidence did not exist! The target–subordinates were randomly assigned to their "highly creative" or "low creative" title.

In addition, Harris, Lightner, and Manolis (1998) found evidence for behavioral confirmation or the self-fulfilling prophecy. Regardless of the awareness conditions, target–subordinates made more suggestions for designing the ad when they were paired with a perceiver–boss who was led to believe the target was highly creative than when paired with a perceiver–boss who was led to believe the target was less creative. The expectations of the perceiver–boss led the target–subordinates to act in a way that would confirm the perceiver–boss' expectations.

The perceiver–boss' awareness of the status differential affected perceptual and behavioral confirmation more than the target–subordinate's awareness. The correlations between the perceiver–boss' expectations and behavior and the target outcomes did not change with target awareness of the status differential. Targets who knew they were subordinates were not more likely to act consistently with their boss' expectations than targets who did not know they were subordinates. Thus, Harris, Lightner, and Manolis (1998) did not find support for the target deference hypothesis.

However, perceivers who knew they were the boss more than those who were unaware of the status differential rated the highly creative target as more creative than the low creative target (perceptual confirmation) and were likely to have targets that acted in ways that were consistent with their (the perceiver-boss') expectations (behavioral confirmation). These findings are consistent with the cognitive economy hypothesis. In an effort to save cognitive energy, bosses do not pay close attention to the behaviors of their subordinates. Rather, they seek out (and even create) evidence that supports their initial expectations. Similarly, Copeland (1993) suggests that motivation (e.g., saving cognitive energy) plays an important role in the self-fulfilling prophecy.

In summary, expectations of subordinates' behaviors drive perceivers to find and create the very behaviors they expect to find. In addition, ambiguous information is often interpreted through the lens of expectation sometimes offering support for the very existence of the expectation (Smith 1998). For this reason, first impressions appear to be validated by the interaction and are, to some extent, self-confirming. This can explain why first impressions tend to be long lasting.

What happens when an expectation is violated? Behaviors that violate expectations (or schemas) are more likely to be noticed and remembered than behaviors consistent with or irrelevant to initial expectations and may play an unduly large role in information processing (Ruble and Stangor 1986; Smith and Graesser 1981). A meta-analysis performed by Stangor and McMillan (1992) revealed that the superior memory for expectation-incongruent information relative to expectation-congruent information was stronger for recall than recognition measures and stronger when perceivers were asked to form impressions rather than when perceivers were asked to memorize information.

7.4.6 How Expectations Affect Interactions with Automated Partners

To what extent do humans' expectations of automated partners drive their human operators to find confirmatory evidence? Given that human operators usually have and are aware of their higher status role in a human–computer "relationship," perceptual confirmation is likely. Human operators probably seek out and find evidence to support their expectations of their automated aids—even when the evidence does not exist. However, automated partners do not typically change their behavior to conform to the human operator's expectations. However, in certain circumstances, such as when the automation actually has the higher status role, it is conceivable that the human might conform to the expectations the automation has of the human. Automation that behaves like a higher status human, or indicates that it has a higher status role, may be more likely to affect the human operator's behavior than automation that behaves like a lower status human.

7.4.6.1 Experiment 4: Violation of Expectations (Dzindolet et al. 2002)

Purpose: What happens when the human expectations of their automated aids are violated? Earlier in this chapter, we presented a study in which participants were provided with very limited information about a task and then were asked to estimate how many errors they and their aid would make (Dzindolet et al. 2002, Study 1). Half the participants were told their aid would be automated (the Contrast Detector); half were told their aid was the prior participant. Participants predicted the automated aids would make about half as many errors as they would make and about half as many errors as the prior participant would make. What happened when these expectations were violated in later trials?

Method: There were 134 participants viewing 200 terrain slides, half of which had a camouflaged soldier. After viewing the slide for three quarters of a second, participants indicated whether or not a soldier was present in the slide. After indicating their absent–present decision, participants were able to view the decision reached by their automated aid, the Contrast Detector. When participants completed all 200 trials, some were told the number of errors they and their aid had made. About one third of the participants were told the automated aid made half as many errors as they; others were told the automated aid made twice as may errors as they. Other participants were not provided any feedback (control).

Finally, students were told they could earn $.50 in coupons to be used at the cafeteria in the Student Union for every correct decision made on 10 randomly chosen trials. Participants had to *choose whether the performance would be based on their decisions or on the decisions of their aid.* After making their choice, students were asked to justify their choice in writing.

Results: Results revealed that participants disused the automated aid. Even among participants provided with feedback that their aid's performance was far superior to their own, the majority (81%) chose to rely on their own decisions rather than on the decisions of the automated aid! Analyses of the justifications of the task allocation decisions provided by participants indicated nearly one quarter (23%) of the participants justified their disuse by stating they did not trust the automated aid as much as they trusted themselves; for example, "The computer didn't earn my complete trust because I swear I saw someone when the computer said there was no one there." Others detailed instances in which the automated aid made an error to justify their disuse. Even among participants told that they made twice as many errors as their automated aid, 57% of the justifications of self-reliance included descriptions of the automated aid's errors. For example, one participant wrote, "I feel that I saw a soldier in a couple of the slides, but the computer said the soldier was not present." Another wrote, "I know for a fact the Contrast Detector was incorrect at least with four slides."

We hypothesize that ironically, the initial high expectations held by the human operator prior to interaction with automated aids may have led to disuse of the automated aid. After limited interaction with the automated decision aid, participants expected the aid to be reliable; they expected it to perform better than average and better than a human aid (e.g., the prior participant), even when warned that the Contrast Detector sometimes makes errors. Because errors made by the automated aid are inconsistent with the human operator's expectation that the automated aid is reliable and accurate, an error made by the automated aid is more likely to be remembered than a correct decision made by the aid, distorting the operator's estimate of the aid's reliability. As the task progressed it may have been hard for the participant to retain an accurate picture of the aid's reliability. This may have led participants to underestimate the performance of the automated aid.

7.4.6.2 Experiment 5: Manipulation of Expectations (Dzindolet et al. 2002)

Purpose: To test these hypotheses, we performed a study in which we manipulated the participants' expectations of their automated decision aid and examined the effect it had on their reliance decisions. To determine the extent to which the manipulation effect would vary based on type of aid, half of the participants were led to believe they were viewing decisions reached by an automated aid (the contrast detector), and half were led to believe they were viewing decisions reached by the prior participant. In reality, participants were provided with the same information.

Method: The procedure used in this study was similar to the one described in the previous experiments. There were 71 Cameron University students completing 200 trials in which they viewed a slide showing a photograph of Fort Sill terrain, determined whether or not a soldier was in the slide, and then were provided with a decision reached by an aid. Half of the participants believed this aid was automated (the Contrast Detector); others believed the aid was a human, the prior participant. All the participants were told that their aid was not perfect.

To manipulate expectations, some of the participants were told that their aid usually made about half as many errors as most participants (positive framing); others were told their aid usually made about 10 errors in 200 trials (negative framing). In reality, the information given to both of these sets of participants was identical, because the average participant working alone makes about 20 errors in the 200 trials. Other participants were not given any specific information about their aid's performance (control).

At the end of 200 trials, participants were told that ten trials would be randomly chosen from the 200 and that $.50 could be earned in coupons for each correct decision. Participants chose whether to rely on their decisions or on the decisions of their aid.

Results: The manipulation of expectations did not affect reliance decisions for participants paired with "human" aids. However, manipulating expectations concerning automated decision aids did lead to more appropriate reliance. Consistent with the idea that violations of expectations are well remembered and play an unduly large role in information processing, participants given information about their automated aid that was framed negatively were more likely to appropriately rely on the superior automated aid than those in the positive or control conditions.

7.4.7 Future Research for HME Claim: Etiquette Is Implicit, Complex, and Implies a Role

In summary, Miller (2002) identified that etiquette is implicit, complex, implies a role, and even conveys intention. We have discussed many of the mechanisms (impression formation, attribution theory, perceptual and behavioral confirmation) by which etiquette plays a role in human–human relationships and extended these principles to examine human–automation "relationships." We found that although these mechanisms occur in both kinds of relationships, the outcomes do vary. Perhaps the most fruitful outcome of this exercise has been the generation of research questions.

7.5 Evaluation of HME Claim: Etiquette Evolves over Time and Conveys Capabilities

As humans get to know one another better, they expect their partners to behave differently in some respects (e.g., self-disclose more, behave with less formality). However, people expect their partners' dispositions and capabilities to remain somewhat stable over time (an honest person is expected to act honestly; a low-performing member is not expected to outperform his or her teammates). It is this stability that makes later interactions more certain and predictable than earlier ones. This predictability is one of the characteristics that is thought to build trust in human relationships. Muir (1987; 1994) extended the interpersonal trust literature to understand human operators' trust of automation and hypothesized that automation that is predictable, dependable, and inspires faith that it will behave as expected in unknown situations will be seen as more trustworthy. Trust is gained in the areas of persistence, technical competence, and fiduciary responsibility.

To test some of Muir's hypotheses, Lee and Moray (1992; 1994) examined the effect of trust in automation on task allocation decisions.

Participants controlled a simulated orange juice pasteurization plant for two hours each day for three days. The simulation included three subsystems, each of which could be operated manually or automatically. Participants could allocate tasks any way they wished and could change their allocations easily. As part of the experiment, whether controlled automatically or manually, one of the subsystems failed periodically. Lee and Moray (1992; 1994) were especially interested in task allocation changes after these failure events, since Muir predicted that after failure events, trust would decline rapidly and slowly increase as the system performed without making errors. At the end of each session, participants completed a questionnaire concerning their trust in the automation and self-confidence in performing the tasks manually.

Results indicated strong individual differences in automation use. Some participants were prone to use manual control; others were prone to use automation. Inconsistent with Muir's hypotheses, the Lee and Moray (1992; 1994) participants rarely changed their allocation task decisions. Once the human operator assigned a subsystem to automated or manual control, he or she was unlikely to change the decision during that session—even after failure events. We found a similar pattern of results using the soldier-detection paradigm. In never before published data, we examined reliance decisions in four blocks of 50 trials each in four different data sets. We found that students remained very consistent in their reliance decisions. In fact, the correlations between the number of reliance decisions made in the first 50 trials with the number made in the last 50 trials were remarkably high (more than .90 in each study).

In an extensive review of the literature on human trust ranging from romantic relationships to work relationships, Lee and See (2004) found that trust had been defined in a variety of ways in different domains. Looking for commonalities among the definitions, Lee and See established that trust has been thought of as an expectation or an attitude, an intention or willingness to act, and a willingness to become vulnerable to the actions of another. In addition, trust usually includes a goal-oriented component. Lee and See (2004) present a dynamic model of trust and reliance on automation that is consistent with Miller's (2002) view that etiquette evolves over time, and they suggest that the way in which the automation interacts with the human operator (e.g., the content and format of the interface) has powerful effects on trust. Similarly, Madhavan and Wiegmann (2007) present a dynamic model of trust. Both models identify the capability or reliability of the automation (and of the human operator) to be important in calibrating trust and affecting automation reliance. The importance of etiquette is an underlying assumption in both of the human–automation models of trust. Both reviews conclude that although trust between humans is comparable or similar in many respects to trust between humans and automation, some differences do exist. For example, Jian, Bisantz, and Drury (2000) find that distrust is more negative when assessing a human than an automated aid.

7.5.1 Future Research for HME Claim: Etiquette Evolves over Time and Conveys Capabilities

Future research should focus on determining the ways in which HME changes over time in human–automated systems and the extent to which these changes affect the human operator's trust in the automation. It is clear that the human factors' models of trust in automation include automation reliability as a key factor. Future research should examine the extent to which HME affects trust in automation and leverage this knowledge to improve human–automation system performance. In addition, research aimed at manipulating the effect of HME on perceptions of automation reliability, thereby affecting trust in automation and automation reliance, should prove to be worthwhile.

7.6 Evaluation of HME Claim: Etiquette Is Different for Different Situations

Miller (2002) suggests there are different etiquettes for different interactions. Just as there are different etiquettes for cocktail parties and work among human relationships, there are different etiquettes for different situations in human–computer relationships. This may lead human operators to expect their automated aids to act differently when the operator is under great stress than when things are going according to plan. Automation that does not change with the situation may be deemed as less reliable and trustworthy.

Although intuitively appealing, little research has been aimed at determining the extent to which HME should differ in different situations, especially when one considers that Miller implies the "different situations" in the above claim are different social situations. Although we are not aware of any studies that specifically address this claim, we did perform a study in which we varied the reliability of the automated aid's decision with the situation (varying the performance situations while keeping the social situation exactly the same).

7.6.1 Variations in Performance Situations Study (Dzindolet, Pierce, Pomranky, Peterson, and Beck 2001)

Purpose: Using a variation of the soldier-detection paradigm, we manipulated the reliability of the automated aid's decision with situation (Dzindolet et al. 2001) in order to determine if participants would be able to vary their automation reliance with automation reliability.

Method: As in past studies, participants viewed slides of Fort Sill terrain, half of which included a camouflaged soldier. Participants were provided with an automated decision aid, the contrast detector, to help them perform their tasks. In this particular study, however, the reliability of the aid varied with soldier presence. Specifically, whenever the soldier was present in the slide, the contrast detector was always correct; it would never miss. However, when the soldier was absent, the automated aid would reach the correct answer only half the time; the aid would operate no better than chance. In other words false alarms and correct rejections were equally likely to occur (see Figure 7.2). However, participants *never* know if the soldier is truly present or absent on a particular slide; they only know the automated aid's decision. Focusing on the automated aid's decision, one can see that whenever the automated aid reached an "absent" decision, the probability of its being wrong was zero. However, when it reached a "present" decision, the probability of it being wrong was .33 (see Figure 7.3).

Half the participants were given explicit training to know the situations in which the aid was likely to be correct and incorrect; other participants were not explicitly given this information. All

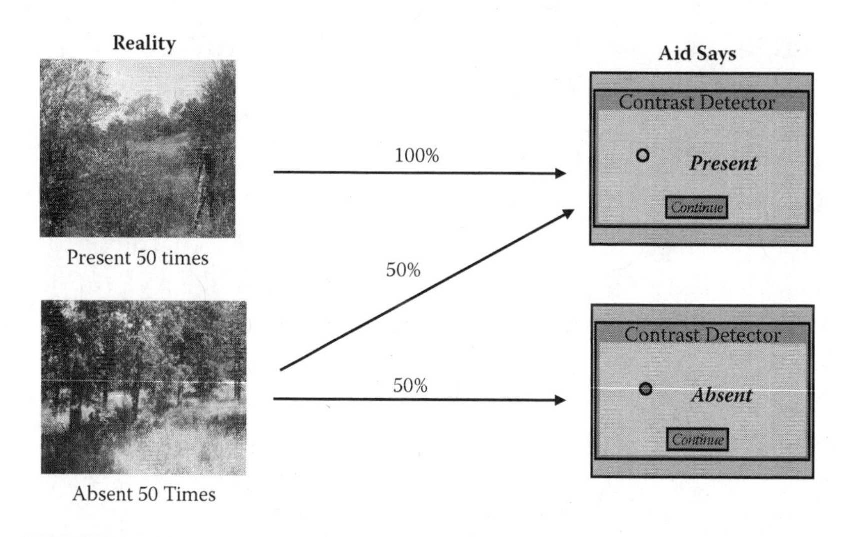

Figure 7.2 Training slides focused on soldier's absence or presence.

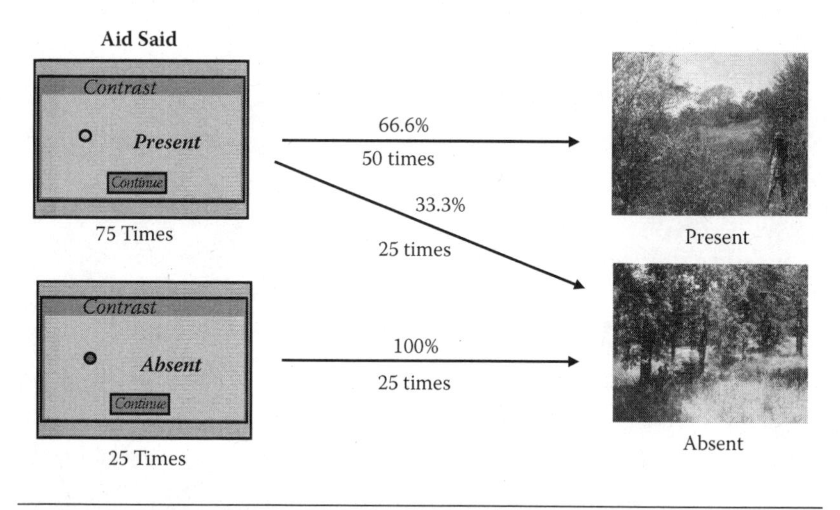

Figure 7.3 Training slides focused on automated aid's decision.

participants were given 200 training slides to "get to know" their automated aids in which the participants gave their absent–present decision *before* viewing the aid's decision. On the 100 key trials, participants viewed the slide for three quarters of a second. Next, they viewed the automated aid's decision. Finally, they made their absent–present decision. Automation reliance was determined by comparing the p(error | aid correct) and the p(error | aid incorrect). Reliance was compared between the situations in which the aid determined the soldier was absent (and the aid was 100% correct) and the situations in which the aid determined the soldier was present (and the aid was only 67% correct).

Results: The results revealed that participants were not sensitive to the reliability of the aid; they were not more likely to rely on the aid when it reached an "absent" than "present" decision. In addition, the explicit training did not improve reliance decisions. The poor reliance decisions reached by the participants led those who had not been provided with an automated aid to outperform those who had! Clearly, participants did not change their expectations of the automated aid with the situation. Perhaps the situations appeared to be too similar to one another for participants to recognize that the etiquette demanded in each situation differed.

7.6.2 Future Research for HME Claim: Etiquette Is Different for Different Situations

Miller's (2002) suggestion that there are different etiquettes for different interactions has much intuitive appeal. However, researchers have yet to assess this HME claim. The fact that human operators do not seem to be sensitive to variations in performance situations does not refute the HME claim that etiquette varies with social situation. After all, the social situation did not vary—only the performance situation. Research needs to be conducted to determine how HME should vary with the social situation in order to maximize human–computer performance.

7.7 Factors Important in Appropriate Automation Reliance

We believe understanding HME will help increase trust in automation and that increased trust in automation will lead to appropriate reliance of humans on automated decision aids and improve human–computer team performance. From the data we have explored in this chapter, what can we conclude?

1. Providing an automated aid does not always increase the performance of the human–automated team.
2. Improving the reliability of an automated decision aid will not necessarily improve the performance of the human–automation team.
3. Although some individual differences exist, with limited information, people expect automated decision aids to perform well and better than human "decision aids."
4. Violations of expectations (e.g., when a human operator notices the automated aid made an error) are likely to be remembered and unduly influence automation reliance decisions.
   a. Reducing initial performance expectations of automated decision aids may lead to more appropriate reliance decisions.
   b. Providing a reason or attribution as to why an automated aid might err may lead to more appropriate reliance decisions.
5. Training of the conditions or situations in which an automated decision aid is (and is not) likely to be reliable is not effective in increasing appropriate reliance.
6. Trust in the aid changes over time and affects automation reliance decisions.

It is clear that our understanding of HME is in its infancy but should prove to be useful to designers and human operators. Table 7.1 reviews the evidence for each HME claim and future research for each HME claim. The pervasiveness and importance of multiple, independent yet interrelated networks to all aspects of life was recognized by the National Research Council (NRC; 2005) in their report on network science commissioned by the U.S. Army. The NRC concluded that the fundamental knowledge necessary to implementing NCO was inadequate. To guide research in NCO, Alberts and Hayes (2006) created a conceptual framework (CF). In the NCO CF, they describe processes in interrelated physical, information, cognitive, and social networks. Inclusion of the cognitive and social networks highlights the importance of individuals, teams, and organizations in NCO. Military research in network science is needed if NCO is to become a reality. Our work focuses on one aspect of the relationship among the four domains by identifying processes and biases

Table 7.1 Review of HME Claims

HME CLAIM	EVALUATION OF HME CLAIM	FUTURE RESEARCH FOR HME CLAIM
Etiquette is implicit, complex, implies a role, and even conveys intention.	• With limited information, people expect automated decision aids to outperform humans. • Counterintuitively, these expectations may lead to reduced trust and reliance on automation if the automation makes a mistake. • Knowing *why* the automated decision aid may err affects trust in automation and automation reliance.	• HME is likely to affect the perceived trustworthiness of the automation, which is likely to affect automation reliance. • HME is likely to affect causal attributions, the explanations people generate to explain why an aid may not act predictably. • HME is likely to affect attribution biases, which should affect trust in the automation and automation reliance. • HME is likely to affect perceptual and behavioral confirmation, which will affect trust in automation and automation reliance.
Etiquette evolves over time and conveys capabilities.	• Current models of trust and automation reliance (e.g., Lee & See 2004; Madhavan & Wiegmann 2007) identify automation reliability to be important in calibrating trust and affecting automation reliance, and hypothesize that trust evolves over time. • HME is likely to be important in calibrating trust (and thereby affecting automation reliance) and should also evolve over time.	• The relationships among HME, automation reliability, trust, and automation reliance need to be explored. • We hypothesize that HME affects trust directly, and that HME can affect perceptions of automation reliability, which will indirectly affect automation trust.
Etiquette is different for different situations.	• This claim has been largely ignored in the human factors literature but has intuitive appeal. • Varying performance situations (keeping the same social situation) did not affect automation reliance.	• This HME claim needs to be examined in laboratory and field studies.

taken primarily from the literature on human–human interactions that may influence automation reliance. Understanding when and why individuals and teams will tend to rely or not rely on automaton is an important factor in designing systems and training programs that promote collaboration among humans and automation. At its core, NCO is about collaboration enabled by and with information age technologies.

References

Alberts, D. and Hayes, R. 2006. *Understanding Command and Control: The Future of Command and Control.* DOD Command and Control Research Program: Library of Congress. ISBN 1-893723-17-8. Pdf file posted on dodccrp.org.

Asch, S.E. 1946. Forming impressions of personality. *Journal of Abnormal and Social Psychology* 41:258–290.

Atkinson, R.K. 2002. Optimizing learning from examples using animated pedagogical agents. *Journal of Educational Psychology* 94:416–427.

Bar, M., Neta, M., and Linz, H. 2006. Very first impressions. *Emotion* 6:269–278.

Berry, D.S. 1990. Taking people at face value: Evidence for the kernel of truth hypothesis. *Social Cognition* 8:343–361.

Buehler, R., Griffin, D., and Ross, M. 1994. Exploring the "planning fallacy": Why people underestimate their task completion times. *Journal of Personality and Social Psychology* 67:366–381.

Copeland, J.T. 1993. Motivational approaches to expectancy confirmation. *Current Directions in Psychological Science* 2:117–121.

Dalal, N.P. and Kasper, G.M. 1994. The design of joint cognitive systems: The effect of cognitive coupling on performance. *International Journal on Human-Computer Studies* 40:677–702.

Darley, J.M. and Fazio, R.H. 1980. Expectancy confirmation processes arising in the social interaction sequence. *American Psychologist* 35:867–881.

Dijkstra, J.J., Liebrand, W.B.G., and Timminga, E. 1998. Persuasiveness of expert systems. *Behaviour and Information Technology* 17:155–163.

Downey, J.L. and Christensen, L. 2006. Belief persistence in impression formation. *North American Journal of Psychology* 8:479–488.

Dzindolet, M.T., Peterson, S.A., Pomranky, R.A., Pierce, L.G., and Beck, H.P. 2003. The role of trust in automation reliance. *International Journal of Human-Computer Studies: Special Issue on Trust and Technology* 58:697–718.

Dzindolet, M.T., Pierce, L.G., Beck, H.P., and Dawe, L.A. 2002. The perceived utility of human and automated aids in a visual detection task. *Human Factors* 44:79–94.

Dzindolet, M.T., Pierce, L.G., Pomranky, R.A., Peterson, S.A. and Beck, H.P. 2001. Automation reliance on a combat identification system. *Proceedings of the Human Factors and Ergonomics Society 45th Annual Meeting*. Santa Monica, CA: Human Factors and Ergonomics Society.

Fiske, S. 1980. Attention and weight in person perception: The impact of negative and extreme behavior. *Journal of Personality and Social Psychology* 38:889–906.

Fullwood, C. 2007. The effect of mediation on impression formation: A comparison of face-to-face and video-mediated conditions. *Applied Ergonomics* 38:267–273.

Hansen, R.D. 1980. Commonsense attribution. *Journal of Personality and Social Psychology* 39:996–1009.

Harris, M.J., Lightner, R.M., and Manolis, C. 1998. Awareness of power as a moderator of expectancy confirmation: Who's the boss around here? *Basic and Applied Social Psychology* 20:220–229.

Heider, F. 1958. *The Psychology of Interpersonal Relations.* New York: John Wiley & Sons.

Helweg-Larsen, M. and Shepperd, J.A. 2001. Do moderators of the optimistic bias affect personal or target risk estimates: A review of the literature. *Personality and Social Psychology Review* 5:74–95.

Holmes, J.G. 1991. Trust and the appraisal process in close relationships. In W.H. Jones and D. Perlman (Eds.), *Advances in personal relationships* (Vol. 2, pp. 57–104). London: Jessica Kingsley.

Jian, J.Y., Bisantz, A.M., and Drury, C.G. 2000. Foundations for an empirically determined scale of trust in automated systems. *International Journal of Cognitive Ergonomics* 4, 53–71.

Jones, E.E. and Davis, K.E. 1965. A theory of correspondent inferences: From acts to dispositions. In L. Berkowitz (Ed.), *Advances in experimental social psychology* (Vol. 1). New York: Academic Press.

Jones, E.E. and Nisbett, R.E. 1972. The actor and the observer: Divergent perceptions of the causes of behavior. In E.E. Jones, D. Kanouse, H.H. Kelley, R.E. Nisbett, S. Valins, and B. Weiner (Eds.), *Attribution: Perceiving the Causes of Behavior* (pp. 79–94). Morristown, NJ: General Learning Press.

Karasawa, K. 1995. An attributional analysis of reactions to negative emotions. *Personality and Social Psychology Bulletin* 21:456–467.

Kelley, H.H. 1950. The warm-cold variable in first impressions of persons. *Journal of Personality* 18:431–439.

Kelley, H.H. 1972. Attribution in social interaction. In E.E. Jones et al. (Eds.), *Attribution: Perceiving the Causes of Behavior.* Morristown, NJ: General Learning Press.

Kelley, H.H. and Michela, J.L. 1980. Attribution theory and research. *Annual Review of Psychology* 31:457–501.

Krueger, J.I., Ham, J.J., and Linford, K.M. 1996. Perceptions of behavioral consistency: Are people aware of the actor-observer effect? *Psychological Science* 7:259–264.

Krull, D.S. and Dill, J.C. 1996. On thinking first and responding fast: Flexibility in social inference processes. *Personality and Social Psychology Bulletin* 22:949–959.

Larzelere, R.E. and Huston, T.L. 1980. The dyadic trust scale: Toward understanding interpersonal trust in close relationships. *Journal of Marriage and the Family* 42:595–604.

Lee, J.D. and Moray, N. 1992. Trust, control strategies and allocation of function in human-machine systems. *Ergonomics* 10:1243–1270.

Lee, J.D. and Moray, N. 1994. Trust, self confidence, and operators' adaptation to automation. *International Journal of Human-Computer Studies* 40:153–184.

Lee, J.D. and See, K.A. 2004. Trust in automation: Designing for appropriate reliance. *Human Factors* 46:50–80.

Lickteig, C.W., Sanders, W.R., Durlach, P.J., Lussier, J.W., Carnahan, T., and Sauer, G. 2003. *A Focus on Battle Command: Human System Integration for Future Command Systems*. Paper presented at the 2003 Interservice/Industry Training, Simulation, and Education Conference, Orlando, FL.

Madhavan, P. and Wiegmann, D.A. 2007. Similarities and differences between human-human and human-automation trust: An integrative review. *Theoretical Issues in Ergonomics Science* 8:277–301.

Merton, R.K. 1948. The self-fulfilling prophecy. *Antioch Review* 8:193–210.

Mezulis, A.H., Abramson, L.Y., Hyde, J.S., and Hankin, B.L. 2004. Is there a universal positivity bias in attributions? A meta-analytic review of individual, developmental, and cultural differences in the self-serving attributional bias. *Psychological Bulletin* 130:711–747.

Miller, C.A. 2004. Human-computer etiquette: Managing expectations with intentional agents. *Communications of the ACM* 47:31–34.

Miller, C.A. 2002. Definitions and dimensions of etiquette. Paper presented at the *AAAI Fall Symposium on Etiquette and Human-Computer Work*, North Falmouth, MA.

Moreno, R., Mayer, R.E., Spires, H.A., and Lester, J.C. 2001. The case for social agency in computer-based teaching: Do students learn more deeply when they interact with animated pedagogical agents? *Cognition and Instruction* 19:177–213.

Muir, B.M. 1987. Trust between human and machines, and the design of decision aids. *International Journal of Man-Machine Studies* 27:527–539.

Muir, B.M. 1994. Trust in automation: Part I. Theoretical issues in the study of trust and human intervention in automated systems. *Ergonomics* 37:1905–1922.

Nass, C., Moon, Y., Fogg, B.J., and Reeves, B. 1995. Can computer personalities be human personalities? *International Journal of Human-Computer Studies* 43:223–239.

National Research Council (NRC) of the National Academies. 2005. *Network Science*. The National Academies Press, Washington, D.C.

Neale, M.A. and Bazerman, M.H. 1985. The effects of framing and negotiator overconfidence on bargaining behaviors and outcomes. *Academy of Management Journal* 28:34–49.

Parasuraman, R. and Miller, C.A. 2004. Trust and etiquette in high-criticality automated systems. *Communications of the ACM* 47:51–55.

Parasuraman, R. and Riley, V. 1997. Humans and automation: Use, misuse, disuse, and abuse. *Human Factors* 39:230–253.

Reeves, B. and Nass, C.I. 1996. *The Media Equation: How People Treat Computers, Television, and New Media Like Real People and Places.* New York: Cambridge University Press.

Rempel, J.K., Holmes, J.G., and Zanna, M.P. 1985. Trust in close relationships. *Journal of Personality and Social Psychology* 49:95–112.

Robins, R.W., Spranca, M.D., and Mendelsohn, G.A. 1996. The actor-observer effect revisited: Effects on individual differences and repeated social interactions on actor and observer attribution. *Journal of Personality and Social Psychology* 71:375–389.

Rosenthal, R. 1994. Interpersonal expectancy effects: A 30-year perspective. *Current Directions in Psychological Science* 3:176–179.

Ross, L. 1977. The intuitive psychologist and his shortcomings: Distortions in the attribution process. In L. Berkowitz (Ed.), *Advances in Experimental Social Psychology* (Vol. 10, pp. 174–221). New York: Academic Press.

Ross, L., Greene, D., and House, P. 1977. The false consensus effect: An egocentric bias in social perception and attributional processes. *Journal of Experimental Social Psychology* 13:279–301.

Ross, L., Lepper, M.R., and Hubbard, M. 1975. Perseverance in self-perception and social perception: Biased attributional processes in the debriefing paradigm. *Journal of Personality and Social Psychology* 32:880–892.

Ruble, D.N. and Stangor, C. 1986. Stalking the elusive schema: Insights from developmental and social-psychological analyses of gender schemas. *Social Cognition* 4:227–261.

Shattuck, L.G. and Miller, N.L. and Miller, G.A. 2006. Using the Dynamic Model of Situated Cognition to Assess Network Centric Warfare in Field Settings. *Proceedings of the 12th International Command and Control Research and Technology Symposium,* Newport, RI.

Singh, I.L., Molloy, R., and Parasuraman, R. 1997. Automation-induced monitoring inefficiency: Role of display location. *International Journal of Human-Computer Studies* 46:17–30.

Skowronski, J.J. and Carlston, D.E. 1987. Social judgment and social memory: The role of cue diagnosticity in negativity, positivity, and extremity biases. *Journal of Personality and Social Psychology* 52:689–699.

Smith, E.R. 1998. Mental representation and memory. In Gilbert, D.T., Fiske, S.T. and Lindzey, G. (Eds.), *The Handbook of Social Psychology (4th ed).* Boston, MA: McGraw-Hill.

Smith, D.A. and Graesser, A.C. 1981. Memory for actions in scripted activities as a function of typicality, retention interval, and retrieval task. *Memory and Cognition* 9:550–559.

Stangor, C. and McMillan, D. 1992. Memory for expectancy-congruent and expectancy-incongruent information: A review of the social and social developmental literatures. *Psychological Bulletin* 111:42–61.

Wicklund, R.A. 1989. The appropriation of ideas. In Paulus, P.B. (Ed.), *Psychology of Group Influence*. Hillsdale, NJ: Lawrence Erlbaum Associates.

Willis, J. and Todorov, A. 2006. First impressions: Making up your mind after a 100-ms exposure to a face. *Psychological Science* 17:592–598.

Wilson, C. 2007. Network Centric Operations: Background and Oversight Issues for Congress. CRS Report for Congress, http://www.crsdocuments.com.

Winston, J.S., Strange, B.A., O'Doherty, J., and Dolan, R.J. 2002. Automatic and intentional brain responses during evaluation of trustworthiness of faces. *Nature Neuroscience* 5:277–283.

8

Etiquette in Distributed Game-Based Training

Communication, Trust, Cohesion

JAMES P. BLISS, JASON P. KRING, AND DONALD R. LAMPTON

Contents

8.1 Introduction

In this chapter we will explore the use of etiquette in training systems derived from commercial video games. The system that will be used as an example is a training and debriefing tool that allows distributed participants to train together in a virtual game environment, and to conduct After Action Review (AAR) sessions in that same environment. In AAR sessions, participants discuss the events that unfolded during training, and how to improve performance in the future. However, we will first present a concise history of human–computer etiquette, followed by a description of the unique etiquette requirements that game systems demand. We will then describe how game systems can be used for team training in complex tasks, and the etiquette requirements that must drive that application. An important consideration we address along the way is the analogy between human–computer and human–human etiquette.

It is clear that computers are at the forefront of almost every human endeavor; commerce, education, entertainment, communication, and transportation are only a few of the areas where computers have changed the nature of tasks. In 1982, John Naisbitt, the author of *Megatrends*, wisely predicted that knowledge of computer systems would be essential for human interaction with the world. In many task domains, human functions are practically subservient to the actions of a computer system. Examples include commercial aviation, nuclear power control, and mass transportation organization and management (Parasuraman and Mouloua, 1996, p. 96).

As humans have gradually adjusted to a world managed in tandem with computer systems, the importance of seamless communication has been demonstrated repeatedly. Numerous real-world anecdotes illustrate the consequences of poor human–computer communication. These include unintended actions automatically performed without human approval, poor task responsibility transfer between humans and computers, communication breakdowns, and a general loss of situation awareness accompanying lack of manual system control.

Many of these issues stem from computer interfaces that do not adhere to humans' expectations for polite and appropriate behavior effective etiquette. Consequently, computer operators may become confused, misled, complacent, or blindly accepting of technology (Singh, Molloy, and Parasuraman, 1993). Poor human–computer interactions and etiquette are particularly evident when machines are used to train complex human behaviors such as teamwork and higher-level cognitive processes like problem solving and decision making. Several recent military training systems, for example, allow multiple human team members to communicate and coordinate actions within a simulated setting, even though many members are not in the same physical location. Through high-speed Internet connections, Voice-over Internet Protocol (VoIP) communication systems, and digital capture and replay capabilities, these systems do much to bridge the distance gap between people and provide the tools for teams to discuss, plan, and improve their performance. Will humans using these systems interact differently than if they were seeing each other face-to-face? Furthermore, some military training systems utilize automated teammates. Human trainees interact with these computer-controlled avatars via text-based or verbal communication to complete mission tasks, such as securing a simulated battlefield. Will humans treat these automated team members with the same level of respect and camaraderie as their human counterparts? And will the avatars be effective?

A specific class of applications on which we will focus in this chapter are computers that mediate human–human communication following computer-mediated, distributed training exercises in the after Action Review Process. We chose this example because it represents a highly complex task with serious consequences and significant potential for etiquette-related problems.

8.2 A Brief History of Human–Computer Etiquette

History has supported the idea that computers have great potential to educate and collaborate with human users. According to Fitts (1951), computers, as machines, do several things better than their human users:

- Computers respond quickly to control input.
- Computers perform repetitive tasks quickly and accurately.
- Computers store and erase information reliably.
- Computers reason computationally.
- Computers perform complex operations rapidly and accurately.
- Computers perform multiple concurrent tasks well.

Yet, situations exist where poor design of the computer interface dialog has led task operators to skeptically prefer manual control, or require that the computer ask for consent before computer decisions are implemented (c.f., Olson and Sarter, 1998).

Some reasons for negative attitudes towards computer control are the interface options available for communication. Historically, humans have interfaced with computer software using languages and metaphors that are restrictive. Though interfaces have grown in sophistication, human users must still interact in ways that cater to computer software limitations. This has made complex tasks like distributed training cumbersome, as trainees must circumnavigate interfaces while learning skills. Game-based training software has improved this situation somewhat. Although errors are possible (and perhaps even likely as game characters require increasingly complex control manipulations by players), the consequences of making errors are not usually as dire, and the potential to learn from errors is more pronounced. An additional benefit is that game users are often more motivated than users of typical software programs (Venkatesh, 1999). Game systems have broad appeal, capturing the attention of children and adults alike. As will be discussed, the game design philosophy is such that users are encouraged to make mistakes and to experiment with interface elements. This flexibility motivates users, and may result in increased learning.

As a result, many instructional software designers are stimulating human learning by incorporating the most promising and beneficial aspects of gaming (Berg, 2000). Such aspects include the freedom to collaborate with others across town, across the nation, or across the world to accomplish some goal. Just as "two heads are better than one" for many tasks, the same assumption is applied to computer-mediated training sessions. However, to benefit from this capability, it is important that team members retain the potential for effective communication during distributed discussion and planning sessions.

8.3 Transfer of Gaming Philosophy to Training Situations

One of the most important issues to be resolved for gaming to be a successful training medium is whether or not it even makes sense to employ games for educational purposes. Off-the-shelf educational software programs for children, such as The Learning Company's Reader Rabbit™ software and Viva Media's Crazy Machines™, have been used successfully to teach children fundamental computer interaction skills, math, spatial relations, and verbal skills for more than a decade. In recent times, there has been much interest in the use of games for educational purposes (c.f., Kirriemuir and McFarlane, 2004).

Yet, the application of gaming technology to the educational realm requires particular attention to elements of each philosophy. Researchers have studied elements of successful task training for many years, with scientists such as Edward Thorndyke, Robert Woodworth,

Charles Osgood, Dennis Holding, Jack Adams, and Irwin Goldstein highlighting central influences on training success:

- Conditions of practice, including similarity of training and test sessions
- Variations of training task and environment fidelity
- Methods for increasing positive task transfer from practice to test
- Distribution of task elements during practice
- Importance of adaptive training and testing

Modern computer software and task training programs stress the importance of maintaining a guided path through training materials, building towers of knowledge systematically by decomposing the target task into elements and requiring mastery of each, one at a time, in lockstep order. Unfortunately, such a progression, and the concomitant presentation of information devoid of context, leads to several disadvantages, including lack of motivation, extended practice periods, inflexible material presentation, adherence to outdated criteria, and predictability of presented material.

Games, by contrast, offer fewer restrictions. Generally speaking, games feature more advanced interaction styles, less predictable task progressions, plot-driven material presentation, tolerance for exploration by learners, a wider range of problem complexity, and the potential for unanticipated solutions.

Another example of the use of games to facilitate learning can be found in Chapter 5. Johnson and Wand describe how they employed game technology to teach language skills. There is evidence that action-based video games may influence some cognitive and sensory processes (Durlach, Kring, and Bowens, 2009; Green and Bavelier, 2003).

Another benefit of game-based training is that it allows for multiple people to interact via the game. Even without traditional face-to-face interactions, in teamed gaming environments, social dynamics often become more pronounced as learners interact. For example, learners must show tolerance for users who make mistakes. Such tolerance reflects a broader emphasis on communication, including trust, cohesion, and other key social features that are present in face-to-face discussions and will be factors in how people interact in computer-mediated discussions like game-based AAR sessions. To better understand how etiquette will develop for the human–machine–human interactions afforded by game-based systems, the following sections briefly discuss communication, trust, and cohesion in the context of distributed AAR and game systems.

8.4 Reconsideration of Classic Communications Models

When two or more individuals combine efforts to accomplish a shared task, communication is central to their performance (Dickinson and McIntyre, 1997; Morgan, Salas, and Glickman, 1993; Prinzo, 1998). Laboratory and field research with small teams and groups shows that when communication is hindered or breaks down, performance suffers. This relationship has been noted in a variety of real-world domains including:

- Medicine (Sexton, Thomas, and Helmreich, 2000)
- Aviation (Kanki and Foushee, 1989; Foushee, 1982; Orasanu, Davison, and Fischer, 1997)
- Human space flight (Cohen, 2000; Kanas, and Caldwell, 2000)
- Emergency response to crises, such as the shootings at Columbine High School in 1996 (Columbine Review Commission, 2001)

In these and other complex domains, communication between team members is the essential element during discussions like an AAR that facilitates collective learning and subsequent task improvement. It is therefore important to consider how computer-mediated interactions, particularly when individuals do not have personal or face-to-face contact, affect the communication process.

Team communication problems happen for many reasons from inadequate verbal skills to excessive amounts of background noise. However, the modality within which a team communicates can have a significant effect on interactions and subsequent performance on shared tasks. In the mid-1970s, Chapanis and his colleagues (Chapanis, 1975) compared traditional face-to-face communication with modes or channels common in that decade: voice only, handwriting, and typewriting. Their work suggested face-to-face communication was superior in most respects because more information could pass between persons and the availability of communication-rich nonverbal cues. Generally, the degree to which communication technologies and modalities feature different levels of "social presence" (Short, Williams, and Christie, 1976) limits interpersonal interaction, particularly for those involved in collective learning processes such as AAR.

Williams (1977) specified three ways in which communication modalities or channels vary. First, channels may differ in terms of bandwidth (number of sensory systems involved in the communication). Face-to-face (FTF) conversations have a higher bandwidth than e-mail messages because individuals can use visual, auditory, and to a lesser extent, tactile cues to convey and receive information. A second way they may vary is the capacity for nonverbal communication. Verbal communication includes vocal cues like inflection, volume, and pacing that are conveyed in spoken communication. Nonvocal communication refers to all other behaviors that are dependent on visual communication, including facial expressions and gaze, body language (e.g., gestures, posture, movement), and touching (Weeks and Chapanis, 1976). The third way communication modalities differ is in their degree of social presence. Williams (1977) pointed out that FTF interactions involve social appraisal of personality, whereas distant media foster comparative impersonal interactions. Several studies (e.g., Stephenson, Ayling, and Rutter, 1976; Wilson and Williams, 1977) suggest audio-only conversations are more depersonalized, argumentative, and narrow in focus than FTF conversations. Although today's game-based training applications have increased social presence, with players choosing avatar styles that match their own appearance or by activating a number of programmed facial gestures to convey emotion, the communication channel they provide is still far less rich than face-to-face interactions.

Modern, distributed human–human teams rely on a large spectrum of communication modes, from two-user communication like chatting and e-mail to group communication via video-teleconferences or multiuser electronic whiteboards. Each mode offers a different level of bandwidth and social presence. Furthermore, serial and parallel forms both have their advantages and disadvantages. Serial communication may diminish the overall quality and quantity of communication because of communication loafing (team members electing not to communicate because others are already doing so).

In contrast, parallel communication, such as voice communication between dozens of teammates in a game-based training exercise, can become very complex and may overload trainees. Multiple, overlapping voice transmissions make it difficult to efficiently understand and share information. Such difficulties illustrate the complex nature of distributed training exercises and AARs where the desire for communication among teammates must be balanced with the need for clarity. A viable alternative is to develop hybrid communication modalities that offer serial and parallel characteristics. Game-based training systems should provide access to a collection of modalities that allows users to select the appropriate channel for a specific purpose. For example, important information that must be broadcast immediately to everyone on the team could be displayed as a bolded text message on the screen that remains visible for a period of time. In other cases, if the leader needs to communicate with only one team member, a directed voice transmission on a unique channel that only the team member can hear would reduce the amount of vocal "clutter."

Game designers sometimes design avatars with a larger set of cues to support more of the properties of face-to-face interactions. Unfortunately, many games still rely on generic avatars that stifle individuation. Accordingly, although team members may be looking at another player while talking over a microphone, almost all of the critical nonverbal communication cues are absent (e.g., raised eyebrows, slumping shoulders, head nods). It is also difficult to tell different team members apart unless additional information, such as the member's name, is visible in the game. Accordingly, organizations using game-based training systems should encourage the development of more life-like avatars, complete with a collection of facial gestures representing basic human emotions and a larger selection of avatar models. Recently, some social networking application environments such as Second Life™ have allowed considerable freedom to manipulate avatar appearance and clothing (see Figure 8.1). The development of more realistic avatars may ultimately enhance other social constructs such as interpersonal trust.

Figure 8.1 Picture of the avatar manipulation controls within Second Life™.

8.5 Trust in Distributed Teaming

One of the most critical etiquette issues influencing computer use is trust. For humans who must use computers to accomplish critical tasks such as medical procedures, military maneuvers or missions, or factory process management, the ability to trust a human or automated counterpart is necessary for successful task completion. In such cases, the consequences of interacting with an untrustworthy partner may be severe: loss of functionality, unpredictable actions, or even loss of life. The importance of trust as a construct defining human–automation relationships has been highlighted by many authors including Parasuraman and Riley (1997), Lee and See (2004), and Muir (1989).

Humans who use games for training are required to show trust of such systems for several reasons. Parasuraman and Riley (1997) have discussed varieties of trust behavior relevant to automation. The optimal behavior is for human operators to trust automation appropriately; that is, to trust a system that warrants trust, and to not trust a system that has been shown unreliable. However, such trust calibration sometimes fails, leading to circumstances where humans may mistrust or distrust systems inappropriately. The term "distrust" has generally been used to refer to an underabundance of trust, often resulting from past unreliable system behavior (see Lee and See, 2004). In other conditions, humans may trust systems too much.

The application of social (human–human) theories of trust to the human–automation relationship is particularly relevant for training or educational AAR, which require people to interact with other human learners. This is a common paradigm; many off-the-shelf games offer the provision for distributed or online game play. Work by the current authors demonstrates that distributed game-based training labs are becoming increasingly common (Lampton et al., 2008). Such labs can provide an effective training approach for distributed teams, but only if trust is optimized and maintained during initial learning, and post-exercise AAR. Adams, Bryan, and Webb (2001) exhaustively discuss the notion of trust in military teams. They pointed out that such teams often depend on both person-based trust and category-based trust. Person-based trust relates to trust of human partners that develops over time, based on experience (Rempel, Holmes, and Zanna's, 1985). Lewicki and Bunker (1996) extend Rempel et al.'s model of person-based trust to apply to work relationships, and thus concentrate less on intense emotional situations and more on information exchange. They propose that trust development follows three stages. The first, calculus-based trust, involves people evaluating trustworthiness of a partner according to costs and benefits expended and received. Calculus-based trust progresses to knowledge-based trust, where task performers form ideas about predictability and dependability because of interactions with the other person in a work context. Identification-based trust, the last stage, depends on one's identification with the motivations, values, and expectations of the other person.

Lewicki and Bunker's model may be most relevant to AAR situations, where task-based information exchange forms the basis for overall learning. If it is true that trust of human teammates and automated game systems progresses in a similar manner, game designers should be encouraged to design interfaces to emphasize trust building elements. For example, participants who are teamed together for the first time may benefit from access to indices of performance that enable calculus-based trust. These could include running score displays, asset lists for other players, and estimates of team member proficiency. As experience in the game progresses, the automated game system could enable development of knowledge-based trust by offering historical displays to review past system and teammate behaviors. Such displays might be accessible in real time, or upon requisition by the player. As trust continues to evolve, it could be useful for the game to portray players (through avatar stereotype or icon) as reflective of a certain behavior type (e.g., an image of a professor for "risk averse and knowledgeable," or an image of a military general as "confident and collaborative"). Such an approach may enable players to better adopt the value systems of other players.

8.6 Cohesion in Distributed Teaming

In addition to trust, another important team variable is the degree to which individual members are devoted to each other and the team's task. In the team performance literature this dedication is termed *cohesion* and is generally broken into two types. *Social cohesion* refers to each person's level of interpersonal commitment to each other, whereas *task cohesion* describes each person's commitment to a team's shared task or goals. In general, cohesion is positively related to team performance (e.g., Miesing and Preble 1985; Strupp and Hausman 1953), but the effect often depends on the type of cohesion (task or social) and the type of task. For instance, on additive tasks which require individual team members to work alone and then combine their efforts to complete an overall group task, Zaccaro and Lowe (1988) found that social cohesion had little effect but high task cohesion increased performance. This was partly explained by their finding that high social cohesion increased nonrelevant conversations that may have cancelled out any positive benefits to performance. However, for disjunctive-type tasks, in which team members must work together to complete a shared task, high levels of both social and task cohesion are associated with better performing teams (Miesing and Preble, 1985; Zaccaro and McCoy, 1988). Furthermore, there is disagreement as to whether high levels of cohesion are a precursor to successful team performance (Bakeman and Helmreich, 1975; Carron and Ball, 1977) or an outcome.

Overall, team cohesion appears to play an important role in how well the team interacts and performs. In the context of game-based training, how team members interact is qualitatively different than in traditional FTF interactions. For instance, members of online game teams, or "clans," rarely, if ever, meet FTF yet have seemingly high levels of individual commitment to performing well in the game and supporting fellow players. These distributed teams "meet" several times a week to practice, members know each other's strengths and weaknesses, and they are able to develop a sense of camaraderie and respect for one another without ever having met personally. Researchers have for years pondered whether and how interpersonal and common goal commitment might develop if members never meet in the same room. In FTF interactions, cohesion partly develops by team members sharing information about themselves. Salisbury, Carte, and Chidambaram (2006) note these exchanges often include nontask-related communications such as likes and dislikes, upcoming vacation plans, even jokes. Can similar exchanges occur via voice- or text-based messages common to game-based training systems? Another important consideration is the degree to which human team members can develop cohesion with automated teammates. In game-based simulations where human users cooperate and complete tasks with computer-driven avatars, often called "bots" in gaming parlance, can humans overcome the fact that their teammates are not real and form some type of interpersonal bond?

To address these issues, it is first necessary to identify a valid measure of cohesion for use with distributed teams. Most available cohesion measures were developed and validated with FTF teams. The Group Environment Questionnaire (GEQ: Brawley, Carron, and Widmeyer, 1987), for example, was initially designed to capture task and social cohesion with sports teams, but its validity outside this context is unclear. Similarly, the Platoon Cohesion Index (Siebold and Kelly, 1988) was designed specifically for U.S. Army soldiers. A third measure, the Perceived Cohesion Scale (PCS: Bollen and Hoyle, 1990), is a six-item measure designed to capture perceptions of belonging and morale. Although originally developed to measure cohesion in larger populations, such as in a city or college setting, Salisbury and colleagues (Chin, Salisbury, Pearson, and Stollak, 1999; Salisbury, Carte, and Chidambaram, 2006) tested whether the PCS would be a reliable and valid measure for smaller groups in two studies for which small work groups used an electronic meeting system to complete a decision-making task. In the first study (Chin et al.), work groups comprised of four to five members interacted in a face-to-face manner while using the electronic meeting system. In the second study (Salisbury et al.), five-person groups again used an electronic meeting system (i.e., a Web-based groupware tool) to communicate and complete a semester-long project; however there were no face-to-face interactions. Results from both studies indicated the PCS had similar psychometric properties when used on these smaller groups. Furthermore, the scale appeared valid for both the face-to-face and virtual groups, suggesting the PCS could be a useful tool in measuring cohesion differences between distributed and collocated groups.

The next step in understanding how cohesion develops in game-based training, particularly between humans and bots, is to conduct a series of comparisons using a relatively unobtrusive measure like the PCS. An initial study should compare cohesion levels between human–human participants in a game and the human–bot relationship. Results would inform efforts to design communication modalities and other interfaces that foster interpersonal exchanges that underlie social cohesion levels, such as the ability for teammates to share personal information. Interface design could also support task cohesion by allowing all team members to view task-relevant information, such as mission status, on their screens during a training exercise. This information may facilitate the development of a shared understanding of the team's goals.

8.7 New Measures of Human Training Performance Focusing on Etiquette

This chapter raises a number of questions as to how game-based training systems, specifically those with the capability to connect distributed team members during AARs, influence human–human and human–computer etiquette, but how do we study these issues? Are extant measures of training performance and team interactions valid for game-based systems or do we need a new set of tools to capture and understand etiquette in this context? For example, communication is central to training and the AAR process and also provides an indication of etiquette patterns for both human–human and human–computer interactions. However, analyzing communication in game systems presents several challenges. One approach is to measure the amount of communication, such as the number of statements exchanged between team members. Several researchers (Jentsch, Sellin-Wolters, Bowers, and Salas, 1995; Mosier and Chidester, 1991) examined the relationship between team performance and the amount of communication during team activities but reported mixed results.

A more promising approach is to identify specific patterns and types of communications. Kanki, Lozito, and Foushee (1989) found that air crews that use speech that is consistent in content and speaker sequence during flight operations outperform teams whose speech is absent of these qualities. Similarly, Bowers, Jentsch, Salas, and Braun (1998) revealed patterns of communication that were indicative of better performance in simulated flight tasks. They analyzed two-statement communication sequences exchanged within good and poor teams. They found that poor teams followed a lower proportion of communications with acknowledgments. These poorer-performing teams also engaged in a higher proportion of nontask related communications.

One of the difficulties of communication analyses is they are time-consuming, especially for team communications. Traditional analyses require researchers to record voice communications during one or more team task scenarios, transcribe the audio recordings into text, and then perform analyses such as visually counting statements, categorizing statements into several basic categories, or, as noted above, counting statement sequences. For this reason, presumably, there are relatively few published studies on team communication analyses. What is needed is an automated, or at least semiautomated, system for analyzing team communications. Fortunately, researchers have proposed several promising techniques in recent years that afford more efficient, objective, and informative communication data collection.

One technique is to analyze the context and usage of team communications by applying a statistical model to words spoken by team members. For example, the process of latent semantic analysis (LSA) uses a mathematical model similar to neural net models to analyze a large set of text data, known as a *corpus* in linguistics (Landauer, Foltz, and Laham, 1998). LSA provides measures of the relationships between words, such as word–word or passage–passage relations, to show meaning based on their association or semantic similarity. A difficulty with LSA is that it requires manual transcription of team communications. However, currently available voice recognition software can be imbedded in game-based simulations to capture communications and facilitate analysis. For example, the latest version of Dragon Naturally Speaking, published by Nuance, has the potential for over 95% accuracy and can capture upwards of 160 words per minute. Furthermore, output is compatible with common programs like Microsoft Word and Excel.

Assuming that efficient communication analysis is possible for game-based simulations, researchers should consider the possible benefit for improving team interactions in terms of etiquette. At one level, a post hoc analysis of a training mission could reveal instances and patterns of negative comments, aggression, or other maladaptive utterances and then relate these instances to objective measures of team performance. For instance, raters may tag the following exchange as negative:

Team Member 1: "Where do we go next?"
Team Member 2: "Don't you remember? I've told you three times we need to take a left!"
Team Member 1: "I know but I forgot."
Team Member 2: "Forget it. Just get out of my way!"

At another level, previous examples, like the one above, could be used to train game participants to recognize ineffective team etiquette. Research by Savicki and colleagues (e.g., Savicki, Merle, and Benjamin, 2002; Savicki, Merle, and Oesterreich, 1998) suggests that training people to use proper etiquette in computer-mediated communication such as e-mail is successful in increasing certain types of communication patterns that emphasize self-disclosure, statements of personal opinion, and "coalition-building" language. Researchers can use similar training approaches to also improve human–computer interactions, highlighting the need for respectful and productive communications.

8.8 Role of Training Facilitators and Interface Designers in Shaping Etiquette

Interactions by military individuals, small teams, and large organizations are guided by long established procedures. A good introduction to military etiquette is presented in Field Manual No. 7-21.13 The Soldier's Guide (Department of the Army, 2003).

The use of distributed, online gaming systems for military training involves the interaction, and potential conflict, of three different types of etiquette: those for computers in general, gaming systems, and in military interactions. The trainees, trainers, and training managers may or may not have significant previous computer or game system experience, but they will be steeped in military etiquette. The optimal approach for developing the human–computer etiquette for online gaming systems for military training will not necessarily require a meticulous recreation of military culture. However, etiquette development should take into account military as well as computer and gaming considerations.

Computer-based training simulations were developed to provide a complement to live training that is less resource intensive, safer, and more flexible. Additionally, game-based training is being actively pursued by the military community: it has the same advantages as simulator-based training, but with much lower equipment and maintenance costs.

Figure 8.2 shows a trainer leading an AAR following a small team exercise conducted in immersive virtual environment simulators. Before the exercise, the trainees met face-to-face for mission planning. During the exercise the trainees each worked in individual, physically separated simulators. They accomplished their team mission via

Figure 8.2 An After Action Review session following an immersive exercise.

individually controlled avatars. During the mission role players and computer generated forces (CGF) represented lateral friendly units, friendly, neutral, or hostile civilians, and enemy forces. After the exercise, the trainees, exercise controller, role players, and operator(s) of computer-generated forces (CGF) met face-to-face to discuss what happened, why events unfolded as they did, and how they could do better the next time.

When participants are collocated, multiple channels of information contribute to an effective and efficient meeting. It is immediately apparent when everyone is back from their break, if anyone is upset about what happened during the exercise, and the rank of the person speaking. The pointer can be passed around as appropriate. In contrast, online game exercises enable trainees, trainers, role players, and CGF operators to be collocated or separated in various combinations. A mechanism must be provided to support the pre-mission exercise planning process. The mission execution phase will most closely parallel simulation-based training. A significant difference will be that bots from the game will be available to supplement the CGF. It is likely that bots will provide a much broader behavioral repertoire than traditional simulation CGF.

The AAR for a distributed exercise presents a number of challenges. Unlike the collocated AAR shown in Figure 8.2, the participants in the distributed exercise will view, hear, and interact from the same stations in which they conducted the mission exercise. The AAR system should provide information about who is speaking (with some indication of their rank and the role they played during the exercise). The pointer will be replaced by a telestrator function operable from each station (Lampton et al., 2008; see Figure 8.3). The etiquette system should referee who has control of the telestrator.

Game-based training systems will not just support training for new leaders but will also provide supplemental training capabilities for experienced leaders (like those shown in Figure 8.2) and established teams. A challenge will be to discriminate among innovations in tactics, and command and control procedures versus unacceptable variations from standard doctrine and procedures. Even though no one can be hurt during game-based training, this in no way eliminates the need to point out mistakes when they are made.

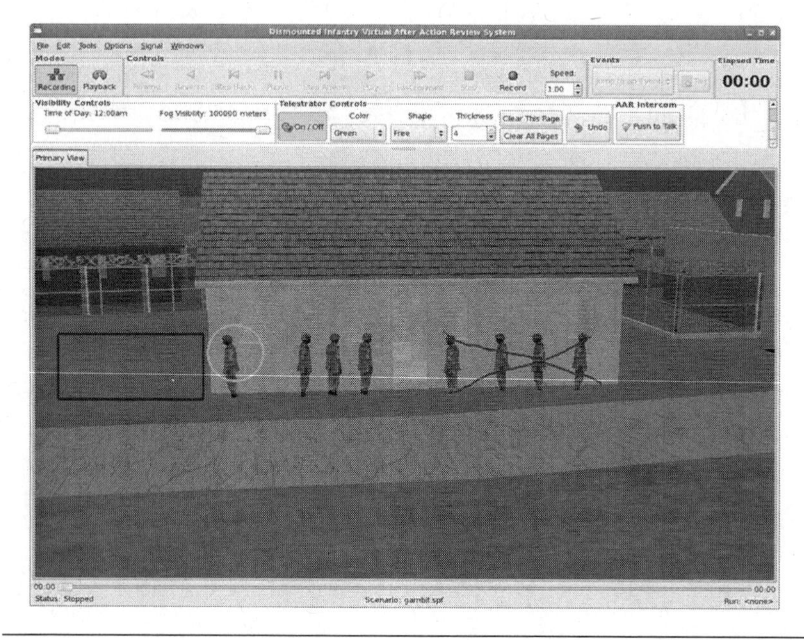

Figure 8.3 Telestrator to support the After Action Review process.

8.9 Recommendations: Etiquette-Based Design of Game-Based Training

It is clear from the preceding discussion that ensuring proper etiquette among technical and human partners is a complex task, particularly when human teammates are distributed, and when technology traditionally used for entertainment is exploited for the purposes of training military duties or other complex tasks. In this chapter, we have attempted to delineate several pertinent influences for successful training in such conditions. The recommendations below are arranged according to the issues already discussed.

Game Philosophy Transfer. It is important to tolerate exploration behaviors, support modification or elaboration of established rules, administer rewards promptly, and tie pedagogical elements to the games' "plot."

Communication. Game designers should be aware of the potential of hybrid communication models, the effectiveness of larger cue sets, and the benefit of well-developed virtual agents.

Interpersonal and Human–Computer Trust. Game designers would do well to facilitate the growth of trust by enabling continuous performance monitoring, asset availability lists, and personality profiling tools.

Interpersonal Cohesion. Distributed game players should be encouraged to establish cohesion with each other by sharing informal anecdotes. Also important for researchers is to develop a usable cohesion measure, and to institute real-time displays of task-relevant information for all participants.

Evaluation of Interaction Success. To determine success of a distributed game-based training session, it is crucial to monitor communication, frequency patterns, and types. It is also important to develop real-time methods for employing advanced statistics for analysis of communication.

The Roles of the Training Facilitator and Interface Designer. For military populations, those who design distributed game-based training should acknowledge the differences in etiquette between conventional teams and those driven by established doctrine. Toward that end, AAR are important to establish conventions and expectations. Equally important are the role expectations for each team member.

References

Adams, B.D., Bryant, D.J., and Webb, R.D.G. (2001). *Trust in teams: Literature review.* (DRDC Toronto Report No. CR-2001-042). Humansystems Incorporated. Guelph, Ontario, Canada.

Bakeman, R. and Helmreich, R. (1975). Cohesiveness and performance: Covariation and causality in an undersea environment. *Journal of Experimental Social Psychology, 11,* 478–489.

Berg, G. A. (2000). Human-computer interaction (HCI) in educational environments: Implications of understanding computers as media. *Journal of Educational Multimedia and Hypermedia, 9*(4), 349–370.

Bollen, K. A. and Hoyle, R. H. (1990). Perceived cohesion: A conceptual and empirical examination. *Social Forces, 69*(2), 479–504.

Bowers, C. A., Jentsch, F., Salas, E., and Braun, C. (1998). Analyzing communication sequences for team training needs assessment. *Human Factors, 40*(4), 672–679.

Brawley, L. R., Carron, A. V., and Widmeyer, W. N. (1987). Assessing the cohesion of teams: Validity of the Group Environment Questionnaire. *Journal of Sport Psychology, 9,* 275–294.

Carron, A. V. and Ball, J. R. (1977). Cause-effect characteristics of cohesiveness and participation motivation in intercollegiate hockey. *International Review of Sport Psychology, 12,* 49–60.

Chapanis, A. (1975). Interactive human communication. *Scientific American, 232*(3), 36–42.

Chin, W. W., Salisbury, W. D., Pearson, A. W., and Stollak, M. J. (1999). Perceived cohesion in small groups: Adapting and testing the Perceived Cohesion Scale in a small group setting. *Small Group Research, 30*(6), 751–766.

Cohen, M. M. (2000). Perception of facial features and face-to-face communications in space. *Aviation, Space, and Environmental Medicine, 71*(9, Section II), A51–A57.

Columbine Review Commission (2001). *The report of Governor Bill Owens' Columbine Review Commission.* Denver, CO: Author (report available: http://www.state.co.us/columbine/).

Department of the Army (2003). *The Soldier's Guide* (Field Manual 7-21.13). Washington, D.C.: Author.

Dickinson, T. L. and McIntyre, R. M. (1997). A conceptual framework for teamwork measurement. In M. T. Brannick, E. Salas, and C. Prince (Eds.), *Team Performance Assessment and Measurement: Theory, Methods, and Applications* (pp. 19–43). Mahwah, NJ: Lawrence Erlbaum.

Durlach, P. J., Kring, J. P., and Bowens, L. D. (2009). Effects of action video game experience on change detection. *Military Psychology, 21*, 24–39,

Fitts, P. M. (1951). Engineering psychology and equipment design. In S. S. Stevens (Ed.), *Handbook of Experimental Psychology* (pp. 1301–1306). New York, NY: Wiley.

Foushee, H. C. (1982). The role of communications, sociopsychological, and personality factors in the maintenance of crew coordination. *Aviation, Space, and Environmental Medicine, 53*, 1062–1066.

Green, C. S. and Bavelier, D. (2003). Action video game modifies visual selective attention. *Nature, 423*, 534–537.

Jentsch, F.G., Sellin-Wolters, S., Bowers, C.A., and Salas, E. (1995). Crew coordination behaviors as predictors of problem detection and decision making times. In *Proceedings of the Human Factors and Ergonomics Society 39th Annual Meeting.* Santa Monica, CA: Human Factors and Ergonomics Society.

Kanas, N. and Caldwell, B. (2000). Summary of research issues in personal, interpersonal, and group dynamics. *Aviation, Space, and Environmental Medicine, 71*(9, Sect. 2), A26–A28.

Kanki, M. A. and Foushee, H. C. (1989). Communication as group performance mediator of aircrew performance. *Aviation, Space, and Environmental Medicine, 60*, 402–410.

Kanki, M. A., Lozito, S., and Foushee, H. C. (1989). Communication indexes of crew coordination. *Aviation, Space, and Environmental Medicine, 60*(1), 56–60.

Kirriemuir, J. and McFarlane, A. (2004). Literature review in games and learning, Futurelab Series, Report No. 8. Bristol: Futurelab.

Lampton, D., Bliss, J., Kring, J. P., Martin, G., Saffold, J., and Garrity, P. (2008, February). *Training Applications of Online Distributed Multiplayer Gaming Systems*. Paper presented at the New Learning Technologies 2008 SALT Conference, Orlando, FL.

Landauer, T. K., Foltz, P. W., and Laham, D. (1998). Introduction to latent semantic analysis. *Discourse Processes, 25*, 259–284.

Lee, J.D. and See, K.A. (2004). Trust in automation: Designing for appropriate reliance. *Human Factors, 46*(1), 50-80.

Lewicki, R. J. and Bunker, B. B. (1996). Developing and maintaining trust in working relationships. In R. M. Kramer and T. R. Tyler (Eds.), Trust in Organizations: Frontiers of Theory and Research. Thousand Oaks, CA: Sage Publications.

Miesing, P. and Preble, J. (1985). Group processes and performance in a complex business simulation. *Small Group Behavior, 16*, 325–338.

Morgan, B. B. Jr., Salas, E., and Glickman, A. S. (1993). An analysis of team evolution and maturation. *Journal of General Psychology, 120*, 277–291.

Mosier, K.L. and Chidester, T.R. (1991). Situation assessment and situation awareness in a team setting. In *Proceedings of the 11th Congress of the International Ergonomics Association*. International Ergonomics Association, Paris, France.

Muir, B. M. (1989). Operators' trust in and use of automatic controllers in a supervisory process control task. Unpublished Doctoral Thesis, University of Toronto, Ontario, Canada.

Olson, W. A. and Sarter, N. B. (1998). "As long as I'm in control ...": Pilot preferences for and experiences with different approaches to automation management. *Proceedings of the 4th Annual Symposium on Human Interaction with Complex Systems*. March 22–25. Dayton, OH.

Orasanu, J., Davison, J., and Fischer, U. (1997). What did he say? Culture and language barriers to efficient communication in global aviation. In R. S. Jensen (Ed.), *Proceedings of the Ninth International Symposium on Aviation Psychology*. Columbus, OH: The Ohio State University.

Parasuraman, R. and Mouloua, M. (1996). *Automation and Human Performance: Theory and Applications*. Mahwah, NJ: Lawrence Erlbaum.

Parasuraman, R. and Riley, V. (1997). Humans and automation: Use, misuse, disuse, abuse. *Human Factors, 39*, 230-253.

Prinzo, O. V. (1998). *An Analysis of Voice Communication in a Simulated Approach Control Environment* (DOT/FAA/AM-97/17). Oklahoma City, OK: Civil Aeromedical Institute, Federal Aviation Administration.

Rempel, J. K., Holmes, J. G., and Zanna, M. P. (1985). Trust in close relationships. *Journal of Personality and Social Psychology, 49*, 95–112.

Salisbury, W. D., Carte, T. A., and Chidambaram, L. (2006). Cohesion in virtual teams: Validating the Perceived Cohesion Scale in a distributed setting. *Database for Advances in Information Systems, 37*(2/3), 147–155.

Savicki, V., Merle, K., and Benjamin, A. (2002). Effects of training on computer-mediated communication in single or mixed gender small task teams. *Computers in Human Behavior, 18*(3), 257–270.

Savicki, V., Merle, K., and Oesterreich, E. (1998). Effects of instructions on computer-mediated communication in single- or mixed-gender small task groups. *Computers in Human Behavior, 14*(1), 163–180.

Sexton, J. B., Thomas, E. J., and Helmreich, R. L. (2000). Error, stress, and teamwork in medicine and aviation: Cross sectional surveys. *British Medical Journal, 320*, 745–749.

Short, J. A., Williams, E., and Christie, B. (1976). *The Social Psychology of Telecommunications*. London: Wiley International.

Siebold, G. L. and Kelly, D. R. (1988). Development of the Platoon Cohesion Index. *US Army Research Institute Technical Report 816*. Alexandria, VA: U.S. Army Research Institute.

Singh, I. L., Molloy, R., and Parasuraman, R. (1993). Automation-induced "complacency": Development of the Complacency-Potential Rating Scale. *International Journal of Aviation Psychology, 3*(2), 111–122.

Stephenson, G. M., Ayling, K., and Rutter, D. R. (1976). The role of visual communication in social exchange. *British Journal of Social and Clinical Psychology, 15*, 113–120.

Strupp, H. H. and Hausman, H. J. (1953). Some correlates of group productivity. *American Psychologist, 8*, 443–444.

Venkatesh, V. (1999). Creation of favorable user perceptions: Exploring the role of intrinsic motivation. *MIS Quarterly, 23*(2), 239–260.

Weeks, G. D. and Chapanis, A. (1976). Cooperative versus conflictive problem solving in three telecommunication modes. *Perceptual and Motor Skills, 42*, 879–917.

Williams, E. (1977). Experimental comparisons of face-to-face and mediated communication: A review. *Psychological Bulletin, 84*, 963–976.

Wilson, C. and Williams, E. (1977). Watergate words: A naturalistic study of media and communication. *Communication Research, 4*, 169–178.

Zaccaro, S. J. and Lowe, C. A. (1988). Cohesiveness and performance on an additive task: Evidence for multidimensionality. *Journal of Social Psychology, 128*(4), 547–558.

Zaccaro, S. J. and McCoy, M. C. (1988). The effects of task and interpersonal cohesiveness on performance of a disjunctive group task. *Journal of Applied Social Psychology, 18*(10), 837–851.

PART IV

Anthropomorphism: Computer Agents that Look or Act Like People

9

Etiquette in Motivational Agents

Engaging Users and Developing Relationships

TIMOTHY BICKMORE

Contents

19th Rule "Let your countenance be pleasant, but in serious matters somewhat grave."

20th Rule "The gestures of the body must be suited to the discourse you are upon."

—George Washington's Rules of Civility*

9.1 Introduction

Micro-analysis of face-to-face interaction among humans can tell us a great deal about how to design etiquette into our human–machine interfaces in order to make them more intuitive, robust, and easy to use. Human protocols for conversational turn-taking, indicating understanding, marking emphasis, and displaying presumed relationship status, are but a few of the myriad verbal and nonverbal cues that can be observed in face-to-face interaction and incorporated into our systems. However, one of the most interesting—and challenging—aspects of human interaction for both analysis and emulation is that the "rules of engagement"—that is, etiquette—expectations change frequently and fluidly during a single encounter. A person must know if their interlocutor is engaging in serious talk, social chat, play, storytelling, or joke-telling to know how to interpret their behavior and to respond appropriately. In addition, these rules can also slowly evolve as a pair of interlocutors develops a relationship over time. This chapter will discuss these phenomena in human social interactions and their ramifications for developing human-like machine interfaces such as the animated health counselor shown in Figure 9.1. This work focuses on a particular type of application as a motivating example—the use of computerized animated counselors for long-term health behavior change interventions—and why at least some of the interaction styles that are effective in this domain comprise rules of etiquette that would not be considered "nice" by Emily Post.

* Washington, George (2003) George Washington's Rules of Civility, fourth revised Collector's Classic edition, Goose Creek Productions, Virginia Beach, Virginia.

Figure 9.1 "Laura," exercise counseling agent.

9.2 Framing in Discourse

The vast majority of dialogue systems developed to date (embodied and otherwise) have been designed to engage people in a strictly collaborative, task-oriented form of conversation in which the communication of factual information is the primary, if not only, concern. As our agents expand their interactional repertoires to include social interaction, play, role-playing, rapport-building, comforting, chastising, encouraging, and other forms of interaction, they will need both explicit ways of representing these kinds of conversation internally and ways of signaling to users that a new kind of interaction has begun. Various social scientists have coined the term *conversational frame* to describe these different forms of interaction.

Gregory Bateson introduced the notion of *frame* in 1955, and showed that no communication could be interpreted without a metamessage about what was going on, that is, what the frame of interaction was (Bateson 1954). He showed that even monkeys exchange signals that allow them to specify when the "play" frame is active so that hostile moves are interpreted in a nonstandard way. Charles Fillmore (1975) defined frame as any system of linguistic choices associated with a scene, where a scene is any kind of coherent segment of human actions (Fillmore 1975). Gumperz (1982) described this phenomena (he called it *contextualization*) as exchanges representative of socioculturally familiar activities, and coined "contextualization cue" as any aspect of the surface form of utterances that can be shown to be functional in the signaling of interpretative frames (Gumperz 1977). Tannen went on to define conversational frames as repositories for sociocultural norms of how to do different types of conversation, such as storytelling, teasing, small talk, or collaborative problem-solving talk (Tannen 1993).

The sociocultural norms take the form of assumptions, scripts (prototypical cases of *what one should do),* and constraints (what one *ought not to do*). These parallel the description of topics that can be taken for granted, reported (talked about; relevant), or excluded, based on sociocultural situations, as described in (Sigman 1983). Scripts can dictate parts of interactions explicitly, as in ritual greetings, describe initiative or turn-taking strategies (e.g., the entry and exit transitions in storytelling and the imperative for the storyteller to hold the floor [Jefferson 1978]), or describe the obligations one has in a given situation, as done in (Traum and Allen 1994) for the collaborative task-talk frame.

The contextualization cues must be used by an agent to indicate an intention to switch frames. While many contextualization cues involve nonverbal behavior (Gumperz 1977), there are also many examples of subtle linguistic cues as well. For example, people rarely say, "Let's do social chat now." Instead, they use ritualized or stereotypical opening moves and topics associated with small talk, such as questions about the weather (Schneider 1988), or phrases indicating the start of a story, for example, "Oh, that reminds me of when ..." (Ervin-Tripp and Kuntay 1997).

9.3 Framing in Relationships

The rules we use for interacting with other people change as our social relationship with them evolves. Similarly, as our computer agents interact with us over longer and longer time periods, it also becomes important to be able to model these relational changes over time. Over the last several years, researchers in social agent interfaces have explored the development of agents that model relational interactions, in which one of the objectives is the establishment of rapport, trust, liking, therapeutic alliance, and other forms of social relationship between a user and an agent (Bickmore and Picard 2005). One of the most important types of behavioral cues in these interactions are those which display social deixis, or *interpersonal stance*, in which one person exhibits his or her presumed social relationship with his or her interlocutor by means of behaviors such as facial displays of emotion, proxemics, and overall gaze and hand gesture frequency.

One way in which language can be used to set relational expectations is through social deixis, or what Svennevig calls "relational contextualization cues" (Svennevig 1999), which are "those aspects of language structure that encode the social identities of participants ... or the social relationship between them, or between one of them and persons and entities referred to" (Levinson 1983). Politeness strategies fall under this general category; facework strategies, for example, behavior to maintain one's social standing in conversation, are partly a function of relationship (Brown and Levinson 1987), but there are many other language phenomena which also fit, including honorifics and forms of address. Various types of relationship can be grammaticalized differently in different languages, including whether the relationship is between the speaker and hearer as referent, between the speaker and hearer when referring to another person or entity, between the speaker and bystanders, or based on type of kinship relation, clan membership, or relative rank (Levinson 1983). One of the most cited examples of this is the *tu/vous* distinction in French and other languages. For example, Laver encoded the rules for forms of address and greeting and parting in English as a (partial) function of the social relationship between the interlocutors, with titles ranging from professional forms ("Dr. Smith") to first names ("Joe") and greetings ranging from a simple "Hello" to the more formal "Good morning," etc. (Laver 1981).

In terms of nonverbal behavior, the most consistent finding in this area is that the use of nonverbal "immediacy behaviors"—close conversational distance, direct body and facial orientation, forward lean, increased and direct gaze, smiling, pleasant facial expressions and facial animation in general, nodding, frequent gesturing and postural openness—projects liking for the other and engagement in the interaction. These behaviors are correlated with increased perception of solidarity, or the perception of "like-mindedness" (Argyle 1988; Richmond and McCroskey 1995). Other nonverbal aspects of "warmth" include kinesic behaviors such as head tilts, bodily relaxation, lack of random movement, open body positions, and postural mirroring and vocal behaviors such as more variation in pitch, amplitude, duration and tempo, reinforcing interjections such as "uh-huh" and "mm-hmmm," greater fluency, warmth, pleasantness, expressiveness, clarity, and smoother turn-taking (Andersen and Guerrero 1998). Research on the verbal and nonverbal cues associated with conversational "rapport" has also been investigated (Cassell, Gill, and Tepper 2007; Tickle-Degnern and Rosenthal 1990).

9.4 Framing in Health Counseling

This research has focused on developing animated conversational agents that can model human social interaction within the context of performing a task, such that the primary purpose of the social interaction is to improve some aspect of task efficacy. In other words, the goal of this work is not just the development of computer friends, but the development of computer agents that are more effective at performing a task *because* they befriend the user first. Within this space, the work has primarily focused on the development of health counseling applications because many studies have demonstrated the importance of trust, therapeutic alliance, and other relational dimensions on health outcomes.

9.4.1 Social Frames

Social dialogue, also known as "small talk," is nearly ubiquitous in professional–client interactions, and health counseling is no exception. Within task interactions, social dialogue is primarily used to establish trust and rapport, but people also use small talk to establish expertise, by relating stories of past successful problem-solving behavior, and to obtain information about the other that can be used indirectly to help achieve task goals. In empirical studies of exercise trainer/client interactions the author found that social dialogue was frequently used, interspersed with talk about the task at hand (Bickmore 2003).

The rules of interaction in a social frame are significantly different than those in a task frame. Small talk is a very ritualized form of interaction in which topics are kept superficial, initially constrained to comments on the interlocutors' immediate surroundings. In addition, agreement with and liking for the other is obligatory, as is demonstration of appreciation for the conversational partner's utterances (Malinowski 1923; Schneider 1988; Cheepen 1988). Stylistically, then, small talk can be seen as a kind of ostensible communication (Clark 1996) in which the interlocutors are pretending to be close friends or acquaintances, while keeping the discourse topics at a safe level of interpersonal distance. This being the case, interlocutors engaged in small talk show signs of positive affect in their speech, and convey many of the signs of "interpersonal warmth," such as close conversational distance, and increased gaze and smiling (Andersen and Guerrero 1998). Structurally, small talk has been characterized (Schneider 1988) in terms of an initial question–response pair, followed by one of several types of third moves (echo question, checkback, acknowledgment, confirmation, evaluation), followed by zero or more synchronized "idling" moves. An example of such an exchange reported by Schneider is:

A: It's a nice morning, isn't it?
B: It's very pleasant.
A: It is really, it's very pleasant, yes.
B: Mhm.

There are also constraints on the introduction of small talk within other types of talk. For example, in conversational frames in which there is an unequal power balance and some level of formality (e.g., job interviews), only the superior may introduce small talk in the medial phase of the encounter (Cheepen 1988). Other style constraints include the increased importance of politeness maxims, and an obligatory turn-taking—one interlocutor cannot hold the floor for the duration of the encounter, and the decreased importance of Grice's maxims such as those described by Grice for promoting "maximally informative communication" (Grice, 1989).

9.4.2 Empathy and Encouragement Frames

The therapeutic alliance (also called "working alliance") has been primarily investigated within psychotherapy for client–therapist relationships, but has also been shown to have a significant impact on outcomes across a wide range of health professions. The therapeutic alliance is defined to be the trust and belief that the counselor and patient have in each other as team members in achieving a desired outcome, and has been hypothesized to be the single common factor underlying the therapeutic benefit of therapies ranging from behavioral and cognitive therapies to psychodynamic therapy (Gelso and Hayes 1998).

While social dialogue is also important for the establishment of therapeutic alliance, the relational factor that is most often mentioned as crucial in forming and maintaining the alliance is the patient's perception of the therapist's empathy for him or her (Gelso and Hayes 1998).

Even in physician–patient interactions, physician empathy for a patient plays a significant role in prescription compliance, and a physician's *lack* of empathy for a patient is the single most frequent source of complaints (Frankel 1995). Based on a meta-analysis of several studies, Buller and Street recommend establishing a "positive relationship" with the client before exercising authority (e.g., giving prescription) (Buller and Street 1992). They also found that physicians who are more expressive, particularly of positive emotion, are more "satisfying" to their patients and that inconsistent or confused emotional expressions by the physician lead to more negative evaluations of them (Buller and Street 1992).

Another important type of interaction in health counseling is encouragement to increase the client's self-efficacy (Bandura 1977). Performatives are explicit statements the coach makes about their belief in the capabilities of their clients to achieve change (e.g., "I really believe you can do it.") (Havens 1986).

9.4.3 Relational Stance in Counseling Interactions

The conversational frames outlined above can change moment-to-moment during the flow of a conversation, marking changes in style of interaction and the rules used to produce and understand utterances. In addition to these very dynamic changes in interactional protocol, much slower changes are typically used to mark evolution in the nature of the interpersonal relationship between the interlocutors. In addition to changes in language, such as increased reference to and reliance on mutual knowledge, increased use of idioms (Bell and Healey 1992), and decreased politeness (Brown and Levinson 1987), a warmer interpersonal stance is often used to stylistically mark a closer relationship, including closer proximity, more frequent gaze, and more smiling and facial animation.

9.4.4 Summary

Based on this literature and the studies of exercise trainer/client interactions, four frames were identified as being particularly important for automated health counseling agents: TASK, for conveying primarily factual information; SOCIAL, for small talk; EMPATHY, for counselor expressions of caring, concern, and empathy for the client; and ENCOURAGE, for counselor performatives, motivational statements, and general encouragement.

9.5 Towards Implementation

In developing automated health counselors with the above capabilities, the primary concern in this work has been with how animated conversational agents can use their bodies and voices to signal changes in conversational and relational frames. This ability is currently implemented within the BEAT text-to-embodied-speech translation system (Cassell, Vilhjálmsson, and Bickmore 2001). BEAT takes the text of an agent's utterance as input (optionally tagged with semantic and pragmatic markers) and produces an animation script as output that can be used to drive the agent's production of the utterance, including not only speech and intonation, but accompanying nonverbal behavior, such as hand gestures, gaze behavior, and eyebrow raises. BEAT was initially developed with a small set of interactional and propositional nonverbal behaviors that it could generate, but was developed to be extensible so that new conversational functions and behaviors could be easily added.

The BEAT system is based on an input-to-output pipeline approach, and uses XML as its primary data structure. Processing is decomposed into modules that operate as XML transducers, each taking an XML object tree as input and producing a modified XML tree as output. The first module in the pipeline operates by reading in XML-tagged text representing the text of the agent's script and converting it into a parse tree. XML provides a natural way to represent information that spans intervals of text, and its use facilitates modularity and extensibility by allowing developers to add their own tags to the parse tree at any stage of processing.

As a simple example, an input to BEAT of the form:

```
<UTTERANCE>It is some kind of virtual actor.</
  UTTERANCE>
```

might result in an output such as the following:

```
<UTTERANCE>
  <GAZE DIRECTION=”AWAY”>It is</GAZE>
  <EYEBROWS DIRECTION=”RAISE”> some
    <GESTURE TYPE=”BEAT”>kind</GESTURE> of
  </EYEBROWS>
  <EYEBROWS DIRECTION=”RAISE”>a virtual
    <GESTURE TYPE=”BEAT”>actor.</GESTURE>
  </EYEBROWS>
</UTTERANCE>
```

This indicates that the agent should look away from the user before speaking begins, look back at the user before the word "some," raise its eyebrows just before the phrases "some kind" and "a virtual," and perform "BEAT" (emphasis) hand gestures just before the words "kind" and "actor."

9.5.1 Relational Stance Frames

As discussed above, one of the most consistent findings in the area of interpersonal attitude is that immediacy behaviors—close conversational distance, direct body and facial orientation, forward lean, increased and direct gaze, smiling, pleasant facial expressions and facial animation in general, nodding, and frequent gesturing—demonstrate warmth and liking for one's interlocutor and engagement in the conversation. BEAT was extended so that these cues would be generated based on whether the agent's attitude towards the user was relatively neutral or relatively warm.

BEAT is designed to over-generate, producing nonverbal behaviors at every point in an utterance that is sanctioned by theory. Thus attitudes conveyed by an animated agent are effected primarily by reducing the number of suggested nonverbal behaviors, as appropriate. For example, in a warm stance (high immediacy), fewer gaze away suggestions are generated, resulting in increased gaze at the interlocutor, whereas, in the neutral stance (low immediacy), fewer facial animation (eyebrow raises and head nods) and hand gesture suggestions are generated. Such cues that are encoded through relative frequency of behavior are currently implemented by means of a StanceManager module which tracks the relational stance for the current utterance being processed, and is consulted by the relevant behavior generators at the time they consider suggesting a new behavior. Centralizing this function in a new module was important for coordination—since attitude (and emotion in general) affects all behaviors systemically. Modifications to baseline BEAT behavior were made at the generation stage rather than the filtering stage, since at least some of the behaviors of interest (e.g., eyebrow raises) are generated in pairs, and it makes no sense to filter out a gaze away suggestion without also filtering out its accompanying gaze towards suggestion.

Relational stance affects not only whether certain nonverbal behaviors occur (i.e., their frequency), but the manner in which they occur. To handle this, the behavior generation module consults the StanceManager at animation compilation time to get a list of modifications that should be applied to the animation to encode manner (the "adverbs" of behavior). Currently, only proximity cues are implemented in this way, by simply mapping the current relational stance to a baseline proximity (camera shot) for the agent; however, in general these modifications should be applied across the board to all aspects of nonverbal behavior and intonation (ultimately using some kind of animation blending, as in Rose, Bodenheimer, and Cohen 1998).

Currently, interpersonal stance is indicated functionally via an attribute in the root-level UTTERANCE tag that specifies what the relational stance is for the current utterance being generated. For example:

```
<UTTERANCE STANCE="WARM">Hi there.</
  UTTERANCE>
```

The generators for gaze, gesture, head nods, and eyebrow movement consult the StanceManager at the time they are about to suggest their respective behaviors, and the StanceManager tells them whether they can proceed with generation or not.

9.5.2 *Conversational Frames*

As mentioned above, people clearly act differently when they are gossiping than when they are conducting a job interview, not only in the content of their speech but in their entire manner, with many of these "contextualization cues" encoded in intonation, facial expression, and other nonverbal behavior.

Contextualization cues are currently implemented in the StanceManager. Conversational frames are marked in the input text using XML tags such as in the following:

<UTTERANCE><EMPATHY>Sorry to hear that you're stressed out.</EMPATHY></UTTERANCE>

During translation of the utterance into "embodied speech," the behavior generation module keeps track of the current frame and when it detects a change in frame it consults the StanceManager for the animation instructions which encode the requisite contextualization cues. The author has implemented four conversational frames for the health counseling agents, based on the empirical studies of human counselor–patient interactions described above: TASK (for information exchange), SOCIAL (for social chat and small talk interactions), EMPATHY (for comforting interactions), and ENCOURAGE (for coaching and motivation).

9.5.3 *Combined Influence*

The interpersonal stance and conversational frame specifications are combined within the StanceManager to yield a final set of modifications to behavior generation and animation modulation, as shown in Table 9.1. Figure 9.2 shows several examples of the effects of stance and frame on proximity and facial expression. For example, in the high immediacy, ENCOURAGE frame condition (lower left cell of

Table 9.1 Effects of Stance and Frame on Nonverbal Behavior

	RELATIONAL STANCE	
FRAME	HIGH IMMEDIACY (WARM)	LOW IMMEDIACY (NEUTRAL)
TASK	Proximity = 0.2 Neutral facial expression Less frequent gaze aways	Proximity = 0.0 Neutral facial expression Less frequent gestures Less frequent head nods Less frequent brow flashes
SOCIAL	Proximity = 0.2 Smiling facial expression Less frequent gaze aways	Proximity = 0.0 Smiling facial expression Less frequent gestures Less frequent head nods Less frequent brow flashes
EMPATHY	Proximity = 1.0 Concerned facial expression Slower speech rate Less frequent gaze aways	Proximity = 0.5 Concerned facial expression Slower speech rate Less frequent gestures Less frequent head nods Less frequent brow flashes
ENCOURAGE	Proximity = 0.5 Smiling facial expression Less frequent gaze aways	Proximity = 0.1 Smiling facial expression Less frequent gestures Less frequent head nods Less frequent brow flashes

Table 9.1) the agent is displayed in a medium shot (halfway between a wide, full body shot and a close-up shot), has a smiling facial expression, and does 50% fewer gaze aways than the default BEAT behavior (thereby spending more time looking at the user).

9.6 Example Application: Relational Agents for Health Counseling

The author has developed a series of health counseling agents that use the above methods for signaling changes in conversational and relational frames. All of these applications are designed to support a series of conversations with the agent for achieving a desired health behavior outcome. In order to model normal trajectories of relationship development, these systems maintain a persistent memory of information about each user and the past conversations the agent has had with them. The work will refer to these agents as "Relational Agents," since they are explicitly designed to establish long-term, social–emotional relationships with their users.

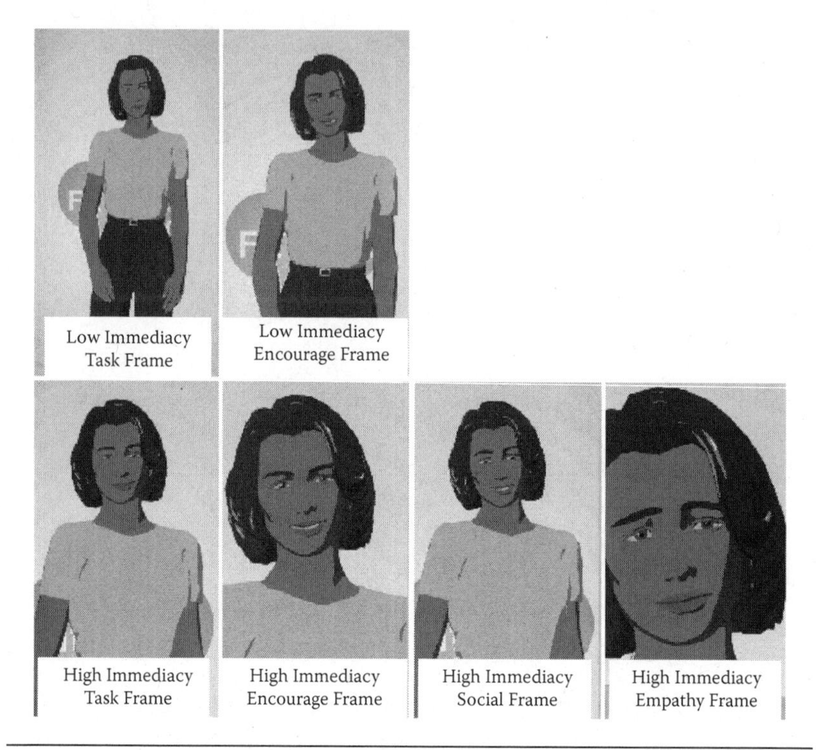

Figure 9.2 Example effects of stance and frame on proximity and facial expression for the "Laura" health counseling agent.

9.6.1 Relational Agents for Exercise Promotion

One of the first relational agents developed for this work was "Laura," the exercise advisor (Figure 9.1). Laura was designed to have daily conversations with sedentary adults on their home computers to attempt to motivate them to do more walking. The interventions were scripted using hierarchical transition networks with agent utterances tagged with conversational frame and relational stance as described previously. This exercise promotion system has been evaluated in two studies to date.

9.6.1.1 Exercise Promotion among MIT Students Laura was first evaluated in a three-arm randomized trial with 101 (mostly) young adults from the MIT community in order to test the efficacy of the agent's relational behavior (Bickmore, Gruber, and Picard 2005). One group of study participants (RELATIONAL) interacted with a version of

Laura in which all of her relational behavior (social dialogue, empathy, nonverbal liking behavior, etc.) was enabled, while a second group interacted with the same agent in which these relational behaviors were removed (NON-RELATIONAL). A third group acted as a nonintervention control and simply recorded their daily physical activity (CONTROL). The Working Alliance Inventory—used to assess the quality of provider–patient relationships in clinical psychotherapy (Horvath and Greenberg 1989)—was used as a primary relational outcome measure.

Participants in the RELATIONAL condition reported significantly higher Working Alliance scores compared with subjects in the NON-RELATIONAL condition, both at one week and at the end of the 4-week intervention. Several other self-reporting and behavioral measures indicated that relational bonding with the agent was significantly greater in the RELATIONAL group compared to the NON-RELATIONAL group. For example, participants in the RELATIONAL group were significantly more likely to use a sentimental farewell with Laura ("Take care, Laura, I'll miss you") rather than a nonsentimental farewell ("Bye") compared to those in the NON-RELATIONAL group at the end of the month ($p = 0.004$). Participants in the RELATIONAL and NON-RELATIONAL groups, combined, increased the number of days per week that they engaged in at least 30 minutes of moderate or more vigorous physical activity significantly more than subjects in the CONTROL condition. However, there were no significant differences between the RELATIONAL and NON-RELATIONAL groups with respect to gains in physical activity.

This study demonstrated that a conversational agent can motivate individuals to do more physical activity over time compared to a group that is simply directed to record their activity. However, it did not demonstrate that relational behavior, when used by such an agent, could boost these results even more. One possible reason for this was the relatively short duration of the study. Subsequent studies have shown that perhaps the biggest advantage of using relational agents over non-relational ones is in maintaining participant engagement with and retention in an intervention, something that may take many months to demonstrate (Bickmore and Schulman 2009).

9.6.1.2 Exercise Promotion among Geriatrics Patients Given the low levels of exercise typically obtained by older adults (only 12% of adults over 75 get the minimum level of physical activity currently recommended by the CDC, and 65% report no leisure time activity (Healthy People 2010, 2000)), the evaluation of the exercise advisor agent within this population seemed critically important. However, there appeared to be significant challenges in getting new technologies such as this accepted and used by this population. Although some researchers have found that many older adults readily accept new technologies such as computers, this segment of the population lags behind all other age groups with respect to computer ownership (only 25.8% of senior households have a computer) and Internet access (14.6% of all senior households have Internet access).*

To evaluate how well the exercise advisor relational agent would work with older adults, the author teamed with geriatricians at the Geriatrics Ambulatory Practice at Boston Medical Center and conducted a pilot test with 21 patients from the clinic (Bickmore et al. 2005). Several modifications were made to the MIT system for this new group of users. The user interface was modified to use large buttons with enlarged text, to allow for easy readability and touch screen input. A numeric keypad (used in conjunction with the touch screen) allowed users to enter their pedometer readings. The system was designed to be used stand-alone—since one could not assume the subjects had Internet connectivity—with subjects provided with a dedicated-use PC, 17" color touch screen monitor (no keyboard or mouse), and table for use during the study. Participants only needed to push the start button on the PC, and it automatically ran the agent interface, conducted a 5–10 minute daily conversation, and then automatically shut down.

The randomized trial compared older adults who interacted with the agent daily in their homes for 2 months (AGENT) with a standard of care control group who were only given pedometers and print materials on the benefits of walking for exercise (CONTROL). Participants ranged in age from 62 to 84, were 86% female, and 73% were African American. There were 17 (77%) who were overweight

* Falling through the Net: Defining the Digital Divide. A report on the telecommunications and information technology gap in America, 1999.

or obese, and 19 (86%) had low reading literacy (Lobach, Hasselblad, and Wildemuth 2003). Eight (36%) had never used a computer before and six (27%) reported having used one "a few times."

All AGENT participants found the system easy to use, rating this an average of 1.9 on a 1 ("easy") to 7 ("difficult") scale. Satisfaction with the overall intervention was very high, with most AGENT participants acknowledging that it was for their benefit: "It was the best thing that happened to me, to have something that pushed me out and get me walking." "I appreciated having that kind of a reminder, because I don't have anybody who will tell me what to do, to remind me, you know, to get up, get out and get some fresh air." Comparisons between the AGENT and CONTROL groups on daily recorded pedometer steps indicated that the slope in the CONTROL group was not significantly different from 0 ($p = 0.295$), while the slope in the AGENT group showed significant increase in steps over time ($p = 0.001$) with this group roughly doubling the number of steps they were walking every day by the end of the study. These results (of agent versus no agent) are consistent with the prior MIT study involving younger adults.

9.6.2 Relational Agents for Medication Adherence

Medication adherence is a huge problem, with many studies indicating that adherence rates of 50% are common, meaning that, on average, patients only take half of the medication they are prescribed as directed (Haynes, McDonald, and Garg 2006). In certain populations, rates of nonadherence are even higher. For example, given the complex regimens that many older adults have, studies have shown that adherence rates of only 40% to 75% are to be expected (Salzman 1995). Another population in which medication adherence is especially troublesome is the group of individuals with mental health conditions, such as schizophrenia. Schizophrenia affects 1% of the population worldwide, and medication adherence to antipsychotic treatments within this population is typically around 50%, leading to higher rates of hospital readmissions and greater number of inpatient days, higher health care costs, and reduced work productivity (Lacro et al. 2002; Dolder et al. 2003).

Leading reasons for medication nonadherence include the stigma of taking medications, adverse drug reactions, forgetfulness, and lack of social support (Hudson et al. 2004). A conversational agent could assist patients in overcoming all of these barriers, including advice to talk their doctors in the case of adverse drug reactions. However, a relational agent could play an especially effective role in both providing social support directly and motivating patients to seek out social support from others.

In collaboration with researchers from the University of Pittsburgh School of Nursing, the authors developed a relational agent-based intervention to promote medication adherence among adults with schizophrenia (Bickmore and Pfeifer 2008). The system runs on a laptop computer as a stand-alone system, and is designed for a 1-month, in-home, daily contact intervention.

The agent tracks each patient's medication taking behavior for a single antipsychotic taken by mouth in pill or capsule form based on self-report, but it also reminds patients to take all of their other medications as prescribed. In addition to medication adherence, the agent promotes physical activity (walking) and talking to the agent every day. For each of these three behaviors, the agent first asks for a self report of behavior, provides feedback on the behavior, and negotiates a behavioral goal. Feedback and goal setting are also provided in summary statements that integrate across the behaviors. For example: "Let's review what you said you were going to do before we chat again. You said you were going to take two doses of Prolixin each day, you were going to try to go for a walk, and you said you would talk to me tomorrow." Intervention on each behavior is started and terminated according to a schedule for the 30-day intervention.

A quasi-experimental pilot study is currently underway to evaluate the medication adherence system, led by researchers at the University of Pittsburgh School of Nursing. There are 20 study participants being recruited from a mental health outpatient clinic who meet the DSM IVR criteria for schizophrenia, are 18–55 years old, are on antipsychotic medication, and have had two or more episodes of nonadherence in the 72 hours prior to recruitment. Study participants were provided with a dedicated use laptop computer for the 30 days of the intervention, as well as Medication Event Monitoring (MEMS) caps to provide an objective measure of medication adherence for the one antipsychotic medication targeted by the intervention.

To date, 10 participants have completed the intervention. System logs indicate that study participants talked to the agent on 65.8% of the available days, with six of the participants talking to the agent at least 25 times during the 30-day intervention. Self-reported medication adherence (gathered through dialogue with the agent; MEMS data is not yet available) was 97%. Self-reported adherence to recommended physical activity (walking) was 89%. Self-reported survey questions on participant attitudes towards the agent indicate that most participants liked and trusted the agent, and 80% of respondents indicated they would have liked to continue working with the agent (Laura) at the end of the 30-day intervention.

9.6.3 Relational Agents for Patient Education

In collaboration with researchers at the Boston Medical Center, the author is currently developing a "virtual nurse" to counsel patients on their self-care regimen before they are sent home from the hospital (Bickmore, Pfeifer, and Jack 2009). A particular focus in this work has been the development of a relational agent that can explain written hospital discharge instructions to patients with low health literacy (Bickmore 2007). To develop this agent, the author videotaped and studied several conversations in which nurses were explaining discharge instructions to patients. From these studies, and many conversations with our collaborators, we developed models of the verbal and nonverbal behavior used by the nurses, and implemented two virtual nurse agents that could emulate much of this behavior (Figure 9.3 shows one them, named "Elizabeth"). In addition to significant knowledge about medications, follow-up appointments, and self-care regimens, the agent was programmed with "bedside manner" (relational behavior) gleaned from the literature and discussions with nurses. This agent will be wheeled up to a patient's hospital bed before they are discharged from the hospital, and spend an hour (on average) reviewing this material with the patient, testing for comprehension, and flagging any unresolved issues for a human nurse to follow up on. This is designed so that patients can have a conversation

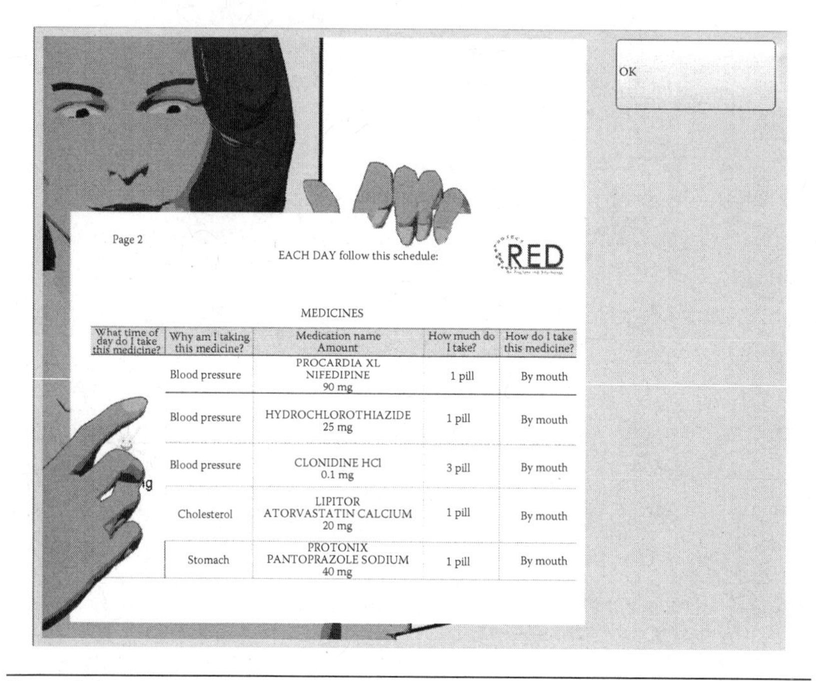

What time of day do I take this medicine?	Why am I taking this medicine?	Medication name Amount	How much do I take?	How do I take this medicine?
	Blood pressure	PROCARDIA XL NIFEDIPINE 90 mg	1 pill	By mouth
	Blood pressure	HYDROCHLOROTHIAZIDE 25 mg	1 pill	By mouth
	Blood pressure	CLONIDINE HCl 0.1 mg	3 pill	By mouth
	Cholesterol	LIPITOR ATORVASTATIN CALCIUM 20 mg	1 pill	By mouth
	Stomach	PROTONIX PANTOPRAZOLE SODIUM 40 mg	1 pill	By mouth

Figure 9.3 "Elizabeth" patient education agent.

with the virtual nurse every day they are in the hospital, so issues of relationship modeling, as well as conversational frame, are important for this application as well.

A randomized clinical trial involving 750 hospital patients is underway. However, as part of the development effort, the author conducted a pilot study to see how well the virtual nurse would do at explaining a discharge document compared to a human explaining the document, for patients with different health literacy levels (evaluated using the REALM instrument (Davis et al. 1993)). There were no differences on post-experiment knowledge tests among the groups. However, lower literacy participants were significantly more satisfied with the agent compared to the human, scored significantly lower on trust in the instructor (whether human or agent), how knowledgeable the instructor was (whether human or agent), and desire to continue working with the instructor (whether human or agent). Dissatisfaction with both human and agent by low literacy patients relative to patients with adequate levels of health literacy likely reflects their overall discomfort discussing a topic they have difficulty with.

Post-experiment interviews revealed that low literacy participants felt the agent exerted less pressure, was less biased, and was more receptive to questions than the human:

> "Elizabeth was cool, I would have taken that again. She was just so clear, she just went page by page so it wasn't missed. And then, I mean you can always just ask them [humans] if you don't understand anyway, but it's different on a screen, I guess, because some people don't want to say that they don't understand. On a screen it's less embarrassing, no one's here so you can say 'Ok, let me hear that again.'"

9.7 Conclusion

Significant progress has been made in understanding how and why people change their rules of interaction (conversational frames), both on a moment-by-moment basis, and in a gradual manner that spans months or years of interaction. The author is now able to replicate much of this behavior in counseling agents, and preliminary evidence indicates that people notice and appreciate when this is done. For example, many study participants commented on Laura's empathic frame, in which she appears close up with a concerned face and depressed speaking style. However, experimentation still needs to be done to determine the degree to which participants notice contextualization cues and the effects they have on participants' attitudes and motivation. Further evaluation is also required to determine if these shifts in conversational frame have a positive effect on counseling outcomes.

The author has now had study participants who have interacted with the exercise counseling agents daily for over 300 days (Bickmore and Schulman 2009), and is developing two health counseling interventions that will last a year in duration. Social behavior, including framing cues and other rules of etiquette, will be crucial in maintaining retention in these programs. When coupled with continually changing social content (e.g., the agent's "back story"), individually tailored and continually updated therapeutic content, and context sensitivity (e.g., shortening the conversation when the patient is in a rush), these techniques could motivate individuals to engage with relational agents for many years before they tire of them.

One interesting outcome of this work is that it clearly exemplifies the two meanings of the term "etiquette"; the first is the more common interpretation meaning polite, considerate behavior. The second interpretation implies normative rules or protocols governing an interaction. Being an effective health counselor often requires confronting clients with information about their behavior that would ordinarily be seen as impolite if said outside of the counseling room. However, such objectivity is part of the expected interactional protocol in health counseling, as is the "drill sergeant" frame, common with many exercise trainers.

As our agents leave the confined roles of information delivery, they will need the ability to signal to users what kinds of interactions they are initiating, and what type of relationship they are expecting to participate in. Richer, more social forms of interaction will enable these agents to function in these roles.

References

Andersen, P. and L. Guerrero. 1998. The bright side of relational communication: Interpersonal warmth as a social emotion. In *Handbook of Communication and Emotion*, edited by P. Andersen and L. Guerrero. New York: Academic Press.

Argyle, M. 1988. *Bodily Communication*. New York: Methuen & Co. Ltd.

Bandura, A. 1977. Self-efficacy: Toward a unifying theory of behavior change. *Psychol Rev* 84:191–214.

Bateson, G. 1954. *A Theory of Play and Fantasy: Steps to an Ecology of Mind*. New York: Ballantine.

Bell, R. and J. Healey. 1992. Idiomatic communication and interpersonal solidarity in friends' relational cultures. *Human-Communication-Research* 18 (3):307–335.

Bickmore, T., L. Caruso, K. Clough-Gorr, and T. Heeren. 2005. "It's just like you talk to a friend"—Relational agents for older adults. *Interacting with Computers* 17 (6):711–735.

Bickmore, T., L. Pfeifer, and Brian W. Jack. 2009. Taking the Time to Care: Empowering Low Health Literacy Hospital Patients with Virtual Nurse Agents. Paper read at *Proceedings of the ACM SIGCHI Conference on Human Factors in Computing Systems (CHI)*, at Boston, MA.

Bickmore, T. and D. Schulman. 2009. A Virtual Laboratory for Studying Long-term Relationships between Humans and Virtual Agents. Paper read at *Proceedings of Autonomous Agents and Multi-Agent Systems*, at Budapest, Hungary.

Bickmore, T. 2003. Relational Agents: Effecting Change through Human-Computer Relationships, Media Arts & Sciences, Massachusetts Institute of Technology, Cambridge, MA.

Bickmore, T. and L. Pfeifer. 2008. Relational Agents for Antipsychotic Medication Adherence. Paper read at *CHI'08 Workshop on Technology in Mental Health*, at Florence, Italy.

Bickmore, T., L. Pfeifer, and M. Paasche-Orlow. 2007. Health document explanation by virtual agents. In *Intelligent Virtual Agents*. Paris.

Bickmore, T. and R. Picard. 2005. Establishing and maintaining long-term human-computer relationships. *ACM Transactions on Computer Human Interaction* 12 (2):293–327.

Bickmore, T., A. Gruber, and R. Picard. 2005. Establishing the computer-patient working alliance in automated health behavior change interventions. *Patient Educ. Couns.* 59 (1):21–30.

Brown, P. and S.C. Levinson. 1987. *Politeness: Some Universals in Language Usage*. Cambridge: Cambridge University Press.

Buller, D. and R. Street. 1992. Physician-patient relationships. In *Application of Nonverbal Behavioral Theories and Research*, edited by R. Feldman. Hillsdale , NJ: Lawrence Erlbaum.

Cassell, J., A. Gill, and P. Tepper. 2007. Coordination in Conversation and Rapport. Paper read at ACL Workshop on Embodied Natural Language, at Prague, CZ.

Cassell, J., H. Vilhjálmsson, and T. Bickmore. 2001. BEAT: The Behavior Expression Animation Toolkit. Paper read at *SIGGRAPH '01*, at Los Angeles, CA.

Cheepen, C. 1988. *The Predictability of Informal Conversation*. New York: Pinter.

Clark, H.H. 1996. *Using Language*. Cambridge: Cambridge University Press.

Davis, T.C., S.W. Long, R.H. Jackson, E.J. Mayeaux, R.B. George, P.W. Murphy, and M.A. Crouch. 1993. Rapid estimate of adult literacy in medicine: A shortened screening instrument. *Fam. Med.* 25 (6):391–5.

Dolder, C.R., J.P. Lacro, S. Leckband, and D.V. Jeste. 2003. Interventions to improve antipsychotic medication adherence: Review of recent literature. *Journal of Clinical Psychopharmacology* 23 (4):389–399.

Ervin-Tripp, S. and A. Kuntay. 1997. The occasioning and structure of conversational stories. In *Conversation: Cognitive, Communicative and Social Perspectives*, edited by T. Givon. Philadelphia: John Benjamins.

Falling through the Net: Defining the Digital Divide. A report on the telecommunications and information technology gap in America. 1999. Washington, DC: National Telecommunication and Information Administration.

Fillmore, C. 1975. Pragmatics and the description of discourse. In *Radical Pragmatics*, edited by P. Cole. New York: Academic Press.

Frankel, R. 1995. Emotion and the physician-patient relationship. *Motivation and Emotion* 19(3):163–173.

Gelso, C. and J. Hayes. 1998. *The Psychotherapy Relationship: Theory, Research and Practice*. New York: John Wiley & Sons.

Grice, Paul. 1989. *Studies in the Way of Words*. Cambridge, MA: Harvard University Press.
Gumperz, J. 1977. Sociocultural knowledge in conversational inference. In *Linguistics and Anthropology*, edited by M. Saville-Troike. Washington, DC: Georgetown University Press.
Havens, L. 1986. *Making Contact: Uses of Language in Psychotherapy*. Cambridge, MA: Harvard University Press.
Haynes, R., H. McDonald, and A. Garg. 2006. Helping patients follow prescribed treatment. *JAMA* 288 (22):2880–83.
Healthy People 2010. 2000. Washington, DC: Office of Disease Prevention and Health Promotion, U.S. Dept. of Health and Human Services.
Horvath, A. and L. Greenberg. 1989. Development and validation of the working alliance inventory. *Journal of Counseling Psychology* 36 (2):223–233.
Hudson, T.J., R.R. Owen, C.R. Thrust, et al. 2004. A pilot study of barriers to medication adherence in schizophrenia. *J. Clin. Psychiatry* 65 (2):211–216.
Jefferson, G. 1978. Sequential aspects of storytelling in conversation. In *Studies in the Organization of Conversational Interaction*, edited by J. Schenkein. New York: Academic Press.
Lacro, J.P., L.B. Dunn, C.R. Dolder, S.B. Leckband, and D.V. Jeste. 2002. Prevalence of risk factors for medication adherence in patients with schizophrenia: A comprehensive review of recent literature. *Journal of Clin. Psychiatry* 63:892–909.
Laver, J. 1981. Linguistic routines and politeness in greeting and parting. In *Conversational Routine*, edited by F. Coulmas. The Hague: Mouton.
Levinson, S.C. 1983. *Pragmatics, Lectures in Linguistics*. Cambridge: Cambridge University Press.
Lobach, D., V. Hasselblad, and B. Wildemuth. 2003. Evaluation of a Tool to Categorize Patients by Reading Literacy and Computer Skill to Facilitate the Computer-Administered Patient Interview. Paper read at *AMIA*, at Washington, DC.
Malinowski, B. 1923. The problem of meaning in primitive languages. In *The Meaning of Meaning*, edited by C.K. Ogden and I.A. Richards: Routledge & Kegan Paul.
Richmond, V. and J. McCroskey. 1995. Immediacy. In *Nonverbal Behavior in Interpersonal Relations*. Boston: Allyn & Bacon.
Rose, C., B. Bodenheimer, and M. Cohen. 1998. Verbs and Adverbs: Multidimensional motion interpolation using radial basis functions. *IEEE Computer Graphics and Applications* 18(5):32–40 (Fall).
Salzman, C. 1995. Medication compliance in the elderly. *J Clinical Psychiatry* 56, Suppl. I:18–22.
Schneider, K.P. 1988. *Small Talk: Analysing Phatic Discourse*. Marburg: Hitzeroth.
Sigman, S.J. 1983. Some multiple constraints placed on conversational topics. In *Conversational Coherence: Form, Structure and Strategy*, edited by R.T. Craig and K. Tracy. Beverly Hills: Sage Publications.
Svennevig, J. 1999. *Getting Acquainted in Conversation*. Philadephia: John Benjamins.

Tannen, D. 1993. What's in a Frame? Surface evidence for underlying expectations. In *Framing in Discourse*, edited by D. Tannen. New York: Oxford University Press.

Tickle-Degnern, L. and R. Rosenthal. 1990. The nature of rapport and its nonverbal correlates. *Psychological Inquiry* 1 (4):285–293.

Traum, D. and J. Allen. 1994. Discourse Obligations in Dialogue Processing. Paper read at ACL '94.

10

Anthropomorphism and Social Robots

Setting Etiquette Expectations

TAO ZHANG, BIWEN ZHU, AND DAVID B. KABER

Contents

10.1 Introduction

Anthropomorphism is the extent to which human-like qualities are attributed to an object or thing by system designers or users. Etiquette is the set of accepted behaviors for a particular type of interaction in a particular group context. In this chapter, we explore the role of anthropomorphism in the design of robots (features and functionality) on human user etiquette expectations. We focus on service robots in hospital environments. A hierarchy of levels of anthropomorphism is presented along with advantages and disadvantages of specific robot features relevant to patient perceptions of social capability.

Underlying factors that may lead to variations in etiquette expectations are also identified. On this basis, a simple design process is proposed to address user etiquette expectations in interactive robot design requiring translation of expectations into anthropomorphic features. This process is constrained by the need to achieve a balance between the illusion of humanness in social robots and the need for certain anthropomorphic features to facilitate particular social functions. This leads to a discussion of measures of effectiveness of anthropomorphic robot features in facilitating perceived etiquette, specifically human emotional responses. General models of human emotion presented in the psychology literature, and various subjective and objective (physiological) measures of emotional states, are identified for assessing human–robot interaction (HRI) design. We conclude that both measures of emotion and traditional measures of HRI effectiveness are critical for evaluating system performance and social capability of interactive robots. Future research directions include quantifying the impact of specific robot interface design feature manipulations on human emotional experiences and etiquette expectations.

10.2 Overview

The objective of this chapter is to identify the relationship between the design of anthropomorphic robots and the etiquette that humans expect to find in interactions with those robots. We propose a design methodology for supporting etiquette in HRI through anthropomorphic designs of robots. We also propose that emotional measures of users can be used to validate the design approach. In the following sections, we attempt to address three research questions related to the objective: (1) Can anthropomorphic design drive etiquette in a social interaction context, and if so why? (2) Can a design process be developed to address etiquette in HRI? (3) Can the effectiveness of anthropomorphic design for etiquette in HRI be measured in service robot applications?

In the following sections, we first define anthropomorphism and discuss its meaning within the scope of designing robots. We examine the relevance of anthropomorphic robot design to human etiquette expectations in human–robot interaction. Through a review of examples of service robots created for health care applications, we identify different degrees of anthropomorphic design and relate them to perceived sociability of robots. We also consider what factors may be relevant to variations in user etiquette expectations. We then propose a five-stage design process addressing etiquette and anthropomorphism in HRI. In the final section, we suggest that measures of human emotion may complement existing task performance measures of HRI to validate the effectiveness of etiquette and anthropomorphic design in service robots.

10.3 Defining Anthropomorphism

Depending on the context, anthropomorphism has several different meanings: (1) the intent and practice of designers to implement human-like features on a nonhuman object (e.g., the Kismet robot designed by Breazeal (2003)); (2) user attribution of human-like qualities to nonhuman organisms or objects, particularly computers and robots (DiSalvo and Gemperle, 2003; Guthrie, 1997; Luczak, Roetting, and Schmidt, 2003; Nass and Moon, 2000); and (3) the attributes of a nonhuman object that may facilitate its work or interaction with humans. These three meanings have been used together in previous literature. For example, Duffy (2003) defined anthropomorphism similar to our second meaning, but discussed anthropomorphism in robots as a design strategy. In this chapter, we focus on people's perception of anthropomorphic features during their interaction with a robot, but also consider it as a design approach when discussing human-like features of a robot.

To explain people's tendency to anthropomorphize nonhuman agents, Gutherie (1997) identified two major theories: (1) People use themselves as models to explain what they do not know in the nonhuman world in the easiest way possible. Anthropomorphic features in robotic technology support this process. (2) People are motivated by emotional needs to form human-like models of nonhuman agents and account for events that may not be logically rationalized or to provide guidance for mitigating certain events, such as errors or failures. For example, people often refer to office automation with personal pronouns (he, she) when a fax machine or copier is not cooperative. Similar to Gutherie (1997), Epley, Waytz, and Cacioppo (2007) theorized that people's tendency to anthropomorphize nonhuman agents is based on three psychological determinants: (1) the accessibility of anthropocentric knowledge for application to agents; (2) the motivation to explain and understand the behavior of other agents (in terms of a human model); and (3) the desire for social contact and affiliation with agents. The second and third determinants are basically identical to the two theses identified by Guthrie (1997). The second determinant is also closely related to Dennett's (1987) intentional stance theory, which argues that viewing the behavior of a system as if it has a mind of its own is generally the best way to guess what that system will do next.

Epley et al. also identified influential factors relevant to all psychological determinants for anthropomorphizing, including dispositional variables (individual differences or personality traits), situational variables, and developmental and cultural variables. Epley et al. said people are more likely to anthropomorphize nonhuman agents when anthropocentric knowledge appears applicable, when they are motivated to be effective social agents, and when they lack a sense of social connection to other agents. Guthrie, however, rejected these determinants and said that anthropomorphism is a common phenomenon (see also Reeves and Nass, 1996), which involves reasonable but erroneous attribution of the most important qualities of humans to parts of the world that do not have them (Guthrie, 1997). That is, perceiving things as human-like may be a class of perceptual error, which results in overestimation in form and function and usually is out of a need for awareness and control. However, the theoretical distinctness of perceptual error (as proposed by Guthrie) and deliberate thought (as proposed by Epley et al.) in anthropomorphizing nonhuman agents may not be so important if one considers the phenomenon as a cognitive process involving both "bottom-up" and "top-down" approaches. Previous studies have provided some convincing evidence that people apply social rules and expectations to computers, based on human–human interaction (HHI), even though they are perfectly aware of the ground truth; that is, they perceive computers to be an equal partner in the interaction, just like another human (Langer, 1992; Nass and Moon, 2000). Consequently, it is likely that the tendency to anthropomorphize technology is always partial in nature. From a practical perspective, what matters most is the effect of anthropomorphic features on people's expectations of social etiquette in interacting with technology and their behavior towards a computer or robot.

10.2.1 Anthropomorphism in (Service) Robot Design

Roboticists have considered human expectations of the shape and manners of things to be critical to design for many years (cf., Corliss and Johnsen, 1968; Jansen and Kress, 1996; Johnsen and Corliss, 1971). In previous telerobotics research—a form of remote robot operation in which a human operator acts as a supervisor (Sheridan, 1989), Johnsen and Corliss (1971) emphasized that the design of robot features should help the operator project his or her presence into a remote work space. They also said that feedback displays should "enhance the operator's identification with the task at hand" (p. 45). On this basis, an early design philosophy for telerobot systems was formed, specifically that anthropomorphic design of manipulator shape/size and spatial correspondence of robot manipulator outputs/movements to hand controller inputs may lead to good overall system performance (Corliss and Johnsen, 1968). However, more recently, Jansen and Kress (1996) demonstrated that physical anthropomorphic design (i.e., human-like shape of robots) may not be particularly important from a performance perspective, but that spatial correspondence of human inputs and robot outputs is strongly related to performance.

Related to Johnsen and Corliss' (1971) early philosophy, for contemporary robots working closely with humans (like service robots) it is a common view that these machines should carry out tasks in a socially acceptable way and communicate with humans to facilitate natural task identification (Dautenhahn et al., 2005; Woods et al., 2007). It has been suggested that the use of anthropomorphic characteristics in robot design (the first meaning of anthropomorphism) may support the sense of social interaction with robots (Fong, Nourbakhsh, and Dautenhahn, 2003; Turkle, 1984). That is, people might interact with robots in a manner similar to how they interact with other humans, as a result of perceptions of anthropomorphic cues from robots.

However, for service robots working autonomously in social settings, several researchers have contemplated whether these machines should mimic human appearance and behavior while attempting to address functional requirements (Breazeal, 2002; Duffy, 2003; Fong et al., 2003). Some survey studies have provided useful information regarding how people generally perceive and internally represent service robots as a basis for design. Khan (1998) described a survey to investigate people's attitudes towards intelligent service robots. In general, he found a strong influence of the science fiction literature. Two important findings for design were: (1) a robot with a machine-like appearance, serious "personality," and round shape is preferred for interactive tasks; and (2) verbal communication using a human-like voice is highly desired. It is interesting to note that subject preferences for physical appearance and vocal capability were not necessarily in agreement relative to addressing anthropomorphic features in robot design. That is, subject preference for physical appearance, relative to the human "prototype," was not as rigid as the preference for vocal function. This is, however, in agreement with Jansen and Kress' observations on telerobot system design (functional anthropomorphic design may play a more critical role in performance than physical anthropomorphic features). Dautenhahn et al. (2005) reported another survey study in which they found a large proportion of participants to view a robot companion as an assistant, machine, or servant. Like Khan (1998), they also found that human-like communication was a desirable trait in a robot companion; whereas, human-like motor behavior and appearance were less essential. A consistent finding from both surveys is that people do not prefer service robots with a highly realistic human appearance but some human-like functionality is considered important for robot effectiveness and efficiency in tasks. These results provide the current design basis for implementing anthropomorphic features in service robots and there is a need for additional research on approaches to anthropomorphic features in service robot design and the perceived implications on task performance.

10.2.2 Levels of Anthropomorphism in Robot Design

In this section, anthropomorphism is referred to as the design practice of implementing human-like features in robots (the first meaning we identified). Based on a review of robotics research, we identified four levels of anthropomorphism in systems design, including: (1) non-anthropomorphic robots; (2) robots designed with some anthropomorphic features as well as intentional deviations from the human "prototype"; (3) robots having an "uncanny" resemblance to the human "prototype"; and (4) highly humanoid robots.

These levels depend on both the robot's physical form (the extent to which the shape resembles a human being) and functionality (the extent to which the actions of the robot mimic those of a human). First, robots at the non-anthropomorphic level are machine-like robots. They lack humanoid features such as facial expression, verbal communication or emotional expressions, and only accept limited input from end users. They are typically considered as autonomous or intelligent machines (Khan, 1998). An example would be a hospital medicine delivery robot that looks like a push-container and seeks delivery confirmation from a nurse with a simple push button interface.

The second level of anthropomorphism, which we refer to as "intentional deviations," includes robots that have certain humanoid features, like two cameras mounted atop the robot giving the appearance of eyes, or a six degree-of-freedom (DOF) gripper giving the appearance of a hand, or a round body shape giving the appearance of human form; however, the robot can be easily distinguished from a real person based on other features, like wheels or tracks, a metal truss structure, etc. In other words, the design of these robots reflects a certain amount of "robot-ness" allowing for quick human identification as a robot, as well as identification of the robot's role in the environment. This level of design can also prevent users from falsely attributing human capabilities to a robot based on the presence of anthropomorphic features. It is important to note that within this level, the more closely a robot resembles the human form, the more empathy and social acceptance it may draw from a human user. We base this on the Epley et al.'s (2007) theory.

The third level of anthropomorphism in robot design is an uncanny resemblance to the human form (Mori, 1982). At this level, users may perceive the robot as "weird." This perception can result from two typical design outcomes: (1) the intent of the designer was to make the robot look and behave like a real human, but subtle imperfections or inconsistencies in the appearance and behavior led to a powerful negative effect on user perceptions; and (2) the intent of the designer was to develop an abstract robot but the ultimate form is perceived as having an uncanny resemblance to the human "prototype." The latter situation can occur if differences between the robot and human form amount to aesthetics only (Bethel and Murphy, 2006; Hanson, 2006).

Lastly, robots with "perfect" anthropomorphic design will presumably be indistinguishable from real human beings, both functionally and physically. Science fiction examples include cyborgs (cybernetic organisms) like "The Terminator," as conceived by James Cameron. Given current robot technology, this may represent an idealized level of anthropomorphism that cannot be achieved. A robot at the fourth level of anthropomorphism would be akin to a computer that can pass the Turing test.

10.2.3 General Advantages and Disadvantages of Anthropomorphism in Robot Design

There is an ongoing discussion over whether or not to use anthropomorphic forms in robot design. In general, using a humanoid design approach can facilitate certain aspects of social interaction, but may also lead to false expectations of robot capabilities and user perceptions of relationships with robots beyond what is actually possible (DiSalvo and Gemperle, 2003). Ideally, as Fong et al. (2003) pointed out, robot design should incorporate anthropomorphic features in such a way as to achieve a balance between illusion of the human "prototype" (to lead the user to believe the robot is sophisticated in areas where the user will not encounter failings) and to support required interactive or social functions (to provide the capabilities necessary for supporting human-like interaction) by robots.

Many researchers favor the idea of adding anthropomorphic features to robots. DiSalvo and Gemperle (2003) contended that anthropomorphism in (robot) design can be useful for four purposes: (1) providing human users with the sense that their interactions with the robot are comparable to human–human interaction; (2) providing users with some basis to mentally explain or understand unfamiliar technologies (see also Goetz, Kiesler, and Powers, 2003); (3) reflecting particular human-like attributes in robots; and (4) reflecting human values. (The second purpose here is similar to Epley et al.'s second psychological determinant of anthropomorphism.) Related to this, the humanoid form has traditionally been seen as an obvious strategy for effectively integrating robots into physical and social environments with humans (Duffy, 2003).

On the other hand, Shneiderman (1989) states that designers employing anthropomorphic features (in human–computer interaction (HCI) applications) may create issues for users including system unpredictability and vagueness—i.e., users may not be certain of what the system is actually capable of. Duffy (2003) comments on Shneiderman's view and says that this situation is not a fault of anthropomorphic features in design, but a fault of designers in not attempting to develop an understanding of people's expectations related to anthropomorphizing the system, and indiscriminately applying certain anthropomorphic qualities to their design. Duffy further suggests that such unmotivated anthropomorphic details may be unnecessary and costly for development. Related to this is Mori's (1982) theory of the "uncanny valley" in perception of anthropomorphic features in which slight inconsistencies in robot design, with respect to people's common expectations, can lead to dramatic negative effects in the interaction with robots, including human disappointment in robot capabilities, misuse of the robot, or disuse. For example, human-like facial features in robots may suggest the capability of facial expression, but the subtleties of human facial expression are difficult to simulate and subsequent user interaction with the robot may lead to negative impressions of actual capability (Duffy, 2003). Referring back to Duffy's argument with Shneiderman's view, it is important that the appearance of a robot be consistent with its behavior and match people's common expectations of a robot in the particular context (Walters et al., 2008). If the actual quality and content of behavior fails to demonstrate that the robot has appropriate human-like intelligence for its job, anthropomorphic features may actually compromise the interaction.

10.2.4 Relevance of Anthropomorphism of Robot Design to Etiquette Expectations

One of the potential uses of anthropomorphic features in robot design is to facilitate social understanding and acceptance of a robot in interactive applications with people in order to achieve defined system goals, just as appearance is important in most human–human interactions. A robot's appearance and behavior provide perceptual cues about its abilities; that is, the design affords (Norman, 1999) the manner of use of the robot. Therefore, differences in robot appearance and associated behavior can actually help users determine how to relate to a particular robot and how to cause the robot to perform desired tasks in human environments (Foner, 1997; Goetz et al., 2003; Walters et al., 2008). For example, Walters et al. (2008) found that, in general, people's expectations of social abilities and reliability of robots are lower for mechanical-looking systems.

More importantly, affordances in robot design may suggest the type of etiquette that is possible in interaction. In general, Bruce, Nourbakhsh, and Simmons (2002) reported that having an expressive face and indicating robot attention to a human through movement make a robot more compelling and inviting to interact with. Luczak et al. (2003) found that people generally treat technical devices with anthropomorphic features as helpers or friends and are more friendly towards such devices than simple tools (see also Hinds, Roberts, and Jones, 2004). They said people often verbalize their emotions, as if talking to the device, to cope with stress during the interaction. Hinds, Roberts, and Jones (2004) studied the effect of robot appearance on people carrying out a joint task with the robot. They said humanoid robots appear to be appropriate for contexts in which people have to delegate responsibility to the robots or when the task is too demanding for the human, and when complacency is not a major concern.

Further, anthropomorphic characteristics may contribute to the elicitation of robot acceptance and human compliance with robot behaviors. Goetz et al. (2003) demonstrated that a robot's appearance and behavior affects people's perceptions of the robot and their willingness to comply with its instructions. Related to this, Burgoon et al. (2000) reported that the amount of anthropomorphism reflected in a computer's design affects its credibility with human users and the extent to which humans consider computer output in decision-making tasks.

Human–human interaction may inform specific etiquette expectations that can be extended to human–robot interaction, dependent upon the degree of anthropomorphism of the robot. It is possible that a robot can use certain social signals to initiate and maintain interaction with a human in a social and comfortable manner. Kanda et al. (2007) investigated human responses to a humanoid robot that expresses cooperative body movements, such as eye contact and nodding (similar to a human listener), in a route guidance situation. Their results showed that the robot displaying cooperative behavior, indicative of interaction with human, received the highest participant evaluations and participant verbal responses to the robot during the study were actually encouraged by the robot's utterances. People appeared to be sensitive to the cooperative behavior displayed by the robot, like cooperative human–human behavior, even though the robot was just pretending to listen to them.

In summary, these prior studies suggest that in tasks involving collaboration between human and robots, or when (service) robots dominate the interaction, anthropomorphic features in robot design may contribute to overall task outcomes and comfort for human users. This notion has been investigated in computer agent design (see the chapter by Bickmore in this book), where "relational agents" designed to establish long-term social relationships with users were found to be effective in health counseling applications.

10.3 Examples of Robots in Applications Requiring Social Interaction

In this section, we discuss several example robots for interactive applications, specifically in the context of hospital working environments. With an aging population and ever increasing number of patients in hospitals, there is a high demand for patient services, such as medicine delivery. Hospital nurses typically perform medicine delivery on a daily basis and this contributes to workload. Consequently, robots have been developed for use in hospitals for such routine tasks, including medicine deliveries, in order to off-load nurses for more critical health care tasks. However, this type of task requires robots to communicate with human users and employees of the hospital.

We initially review three hospital robot systems, including physical and interface features, and how their features affect user etiquette expectations in interacting with the robots. Here, we define users to include patients, nurses, pharmacists, and even robot maintenance personnel. Secondly, we discuss variations in patient etiquette expectations in the course of HRI and underlying factors, including: (1) contextual dynamics (i.e., different operating environments and patient health states); and (2) different degrees of patient anthropomorphizing of robots based on exhibited behaviors. Finally, we subjectively assess the levels of anthropomorphism of several hospital robots and relate this to the degree of social etiquette each robot is capable of.

10.3.1 Nursing/Hospital Service Robots

The Helpmate is a robotic courier designed to perform delivery tasks (see Figure 10.1, Plate A), including carrying meal trays, medications, medical reports, samples, and specimens. At the same time, the Helpmate is designed to exhibit certain human-like behaviors while navigating crowded hospital hallways, including prompting hospital users to its presence (Engelberger, 1998; Evans, 1994; Kochan, 1997; Krishnamurthy and Evans, 1992). The robot is totally machine-like in appearance, with a bumper surrounding the bottom of the robot, a multilevel storage cart as the body, and one on-board keypad mounted atop the robot. The Helpmate is programmed with an internal map of a hospital for autonomous navigation and integrates several sonar and infrared sensors to avoid obstacles. One form of etiquette the Helpmate is capable of is expression of intent (to action). The robot uses flashing lights during navigation to indicate when it is turning. The Helpmate also notifies nurses of its arrival at stations and waits for some time to be attended to before moving on to the next point of delivery in a predefined route.

The Pearl is another hospital-grade robot (see Figure 10.1, Plate B), primarily designed to serve as a cognitive reminder for patients and an intelligent walker (Pollack et al., 2002). It has humanoid features including a movable head with simulated eyes, eyebrows and a mouth,

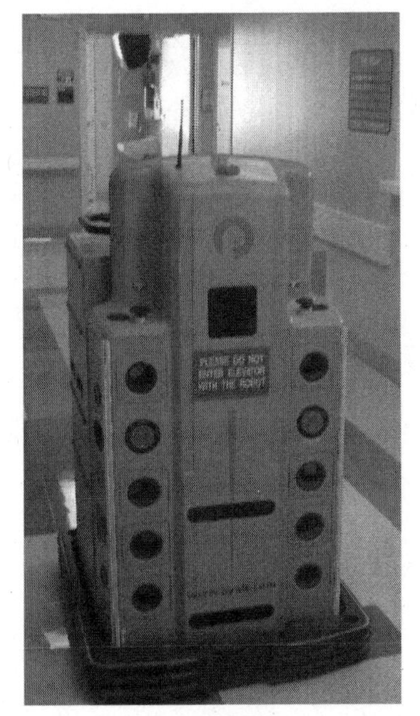

Plate A

Plate B

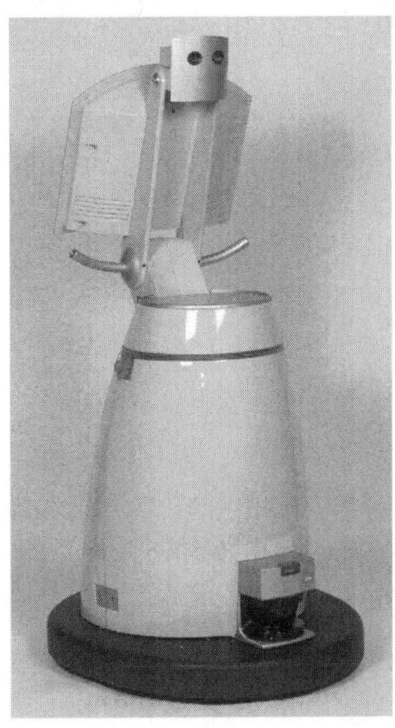

Plate C

Figure 10.1 Images of existing hospital-grade robots. Plate A: Helpmate robot; Plate B: Pearl robot; Plate C: Care-O-bot.

and arms. The Pearl is capable of synthesized speech and has a touch-screen display, located at eye height for a sitting patient, to make input. The features of the robot's face (eyebrows and mouth) are motorized and Pearl is capable of different facial expressions by modifying the angle of its mouth and/or eyebrows. The face can also be swiveled at high speeds and the eyes are motorized to suggest saccadic movement, when tracking a human user's face. The designers of Pearl targeted a set of "polite" behaviors when developing the robot's cognitive support and patient navigation assistance functions, including: (1) the system should be aware of the activities the patient is expected to perform (e.g., talking, remembering, walking); (2) the system should not annoy the patient in interaction; (3) the system should prevent patient over-reliance; and (4) the system should use adaptive instead of fixed pacing in movement to support patient walking. The first three objectives can be considered etiquette constraints for the design process, including being aware of the human user's goal states, being courteous in interacting with humans, and ensuring patients do not become dependent on the robot for needs.

The Care-O-bot is another assistive robot (see Figure 10.1, Plate C) that is used in hospital or home environments (Schaeffer and May, 1999). The robot has a six-DOF manipulator to hold and fetch objects, two cameras simulating eyes, and a liquid crystal touch-screen display mounted on top of the robot for interaction. The cameras and display can also be used for telemedicine (remote medicine) applications, such as a doctor observing and consulting with a patient through the robot interfaces. Beyond this, the Care-O-bot can serve as an intelligent walker, like the Pearl. A user can directly control robot movements or the Care-O-bot can automatically lead the user to a destination point. This functional flexibility helps to prevent users from forming the expectation that the Care-O-bot will dominate the interaction. The highly anthropomorphic features of the robot, including the gripper (hand) and cameras (eyes), also support user perception that the robot is capable of social etiquette, as in HHI.

10.3.2 Variations in User Etiquette Expectations

Two factors that may drive human etiquette expectations during the course of HRI applications include contextual dynamics and changing

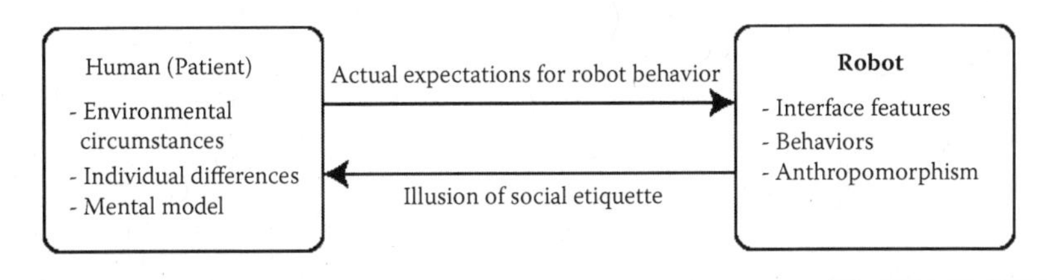

Figure 10.2 Underlying factors in patient–robot interaction.

perceptions of levels of robot anthropomorphism (see Figure 10.2). For example, patients may select different mental models of robot behavior depending upon the location at which they encounter a robot in a hospital (e.g., a room, a nurse's station, a hallway, a waiting room, etc.). This will affect an individual's expectation of the type of interaction they might have with the robot. In hallways, a robot might be expected to verbally communicate when movement is not possible, or in attempting to navigate around obstacles. This is necessary to prevent human confusion and frustration in the interaction. At a nurse's station, a robot may be expected to announce itself, accept input from a nurse to "stop" and relinquish its cargo, and accept input to go on its way.

In addition to environmental conditions dictating user expectations of robot behavior, individual differences may also lead to different etiquette expectations. We conducted an HRI experiment in which a robot with varying levels of anthropomorphic features (abstract or human-like face, synthesized or digitized speech, interaction with the robot through a touch screen or not) was programmed to deliver "medicine" for senior citizens in an assisted-living facility. At the end of the experiment, participants were asked what was the key factor influencing their perception of robot etiquette during the interaction. More than half of the participants (15 out of 24) said that speech communication was the most critical of all features. In addition to speech, a large group of participants (19 out of 24) also liked a face on the robot for suggesting humanness; however, the remaining participants did not think that the face influenced the perception of "humanness." Most of the participants said they would prefer human-like features for robots in hospitals but the robots should provide reassurance of their capability to perform functions suggested by their features (e.g., accuracy and verification of task results). These preliminary results showed that despite general trends in perceptions of robot features, there are individual differences that occur in assessing robot etiquette.

As patients and nurses interact with robots, their mental models of how a robot operates, and how they should respond, may change over time. A patient's original mental model may be based on concepts, beliefs and experiences of HHI and could be expanded or constrained by changes in the robot's appearance or patient experiences of different types of robot behaviors (Kiesler and Goetz, 2002). This may in turn mediate the extent to which the patient anthropomorphizes the robot and expects social interactivity.

10.3.3 Level of Robot Anthropomorphism and Potential for Social Etiquette

In Table 10.1, we list the three health care robots reviewed above and present a general classification of each robot in terms of the levels of anthropomorphism in robot design defined in Section 10.2.2. Both the fidelity of robot physical features, relative to human features, and the fidelity of functional behaviors, relative to human behaviors, are considered in the overall classifications. The classification of physical and functional behaviors, as primary categories, is based on robotics literature. Although there may be other factors influencing anthropomorphism, like the "personality" of robots, they can be grouped under functional behaviors, in a general sense. In other words, we consider functional fidelity to include psychomotor (speech and motion) as well as cognitive abilities of the agent (robot). We subjectively rated the level of social etiquette exhibited by each robot on a scale from "no sociability" to "highly sociable" based on descriptions of each robot available in the literature. This is a very simple approach by which to assess the overall level of anthropomorphism of an interactive robot and is by no means intended to represent a comprehensive evaluation of all factors that may dictate the perceived human-ness of a robot.

Table 10.1 Anthropomorphism and Etiquette of Hospital Delivery Robots

ROBOT	PHYSICAL FIDELITY	FUNCTIONAL FIDELITY	ETIQUETTE
Helpmate	Box like	Simple auditory feedback	Limited sociability
Pearl	Head with eyes and arms	Synthesized speech; adaptive speed	Medium sociability
Care-o-bot	Manipulator as hand; two cameras as eyes; and round shape	Natural human speech; holding and fetching objects; adaptive speed	High sociability

One interesting outcome that can be noted on the subjective ratings is the correlation of the overall level of anthropomorphism with the perceived degree of sociability of the robot. The Care-O-bot has a high level of human-likeness and is capable of a high level of social etiquette. However, the Helpmate is a machine-like delivery robot and has a far lower perceived social etiquette capability. Although this assessment is based strictly on subjective ratings, there is a strong correspondence of the level of anthropomorphism and the perceived sociability of the robots.

10.4 Designing for Etiquette and Anthropomorphism in Human–Robot Interaction (HRI) Applications

In order to achieve a critical balance between the illusion of humanness in a robot and integration of anthropomorphic features to support specific interactive functions for social applications (as suggested by Duffy (2003)), there is a need for a structured robot design approach. Here, we propose a simple five-stage process aimed at achieving appropriate physical and functional anthropomorphism and the potential for social etiquette in interactive applications, as in the hospital domain.

Stage 1—Conduct a contextual analysis: The first step as part of the design process is identification of the user and robot roles in the interaction, as well as the context in which the interaction is to occur. The human may act as a supervisor, partner, or support person to the robot. Depending upon his or her role, the human will select an "appropriate" mental model for interaction with the robot, including their beliefs and concepts of various robot states, as well as procedures or conventions for interaction. The set of conventions, as part of the mental model, may effectively constitute the human's concept of etiquette in interacting with the robot. Therefore, it is important to characterize the user's mental model as a basis for designing the robot. This can be done using concept mapping (Novak and Gowin, 1984) or other types of knowledge elicitation techniques. As previously discussed, the environment in which the user encounters the robot should also be considered, as it may influence mental model selection.

Stage 2—Prepare an inventory of user etiquette expectations: If a user's internal model (schema) of situations in which he or she may encounter a robot can be characterized, along with the action scripts they follow in dealing with the robot, the implicit aspects of etiquette, which are important to the interaction, may be identified. Users should be asked to identify the social and behavioral expectations they have for interaction with the robot through interviews, questionnaires, etc. These should focus on what is considered to be socially attentive, thoughtful, and courteous behavior in the context. Questions should also address what mannerisms (e.g., blinking), speech (e.g., phrases), and action (e.g., gestures) are expected by users. This inventory is intended to make the implicit etiquette expectations of the user explicit as a basis for design.

Stage 3—Define the appropriate level of robot anthropomorphism: This stage involves developing a basic system design to address user etiquette expectations. A designer must translate the elements of the etiquette expectation inventory into necessary anthropomorphic robot features and functions. It is also considered helpful to identify the overall level of anthropomorphism for the robot design. Interface features should be prototyped to address specific interactivity requirements of the user. Each feature and communication channel should be considered for its potential impact on anthropomorphizing and social expectations.

Stage 4—Test and generate outcome measures: In this stage, measures of user response and overall system performance are applied to determine the effectiveness of the prototype robot design for achieving social interaction as well as task goals. Typical HRI performance measures of efficiency and effectiveness should be used to assess whether the robot is capable of achieving its objectives; however, measures of user emotional response may be necessary to provide for assessment of whether etiquette expectations are satisfied by the design. (This is the focus of the final section of this chapter.) In any evaluation of a service or assistant robot intended to support some degree of social interaction, human–human interaction should be considered as a benchmark. (Of course, this would not be a useful standard in service robot applications in which sociability is not a desired feature.) Results of performance and user emotional state assessments in robot tests should be linked to potential violations of human–human interaction convention that may be intrinsic to the HRI design. These violations should provide a basis for improving the prototype design.

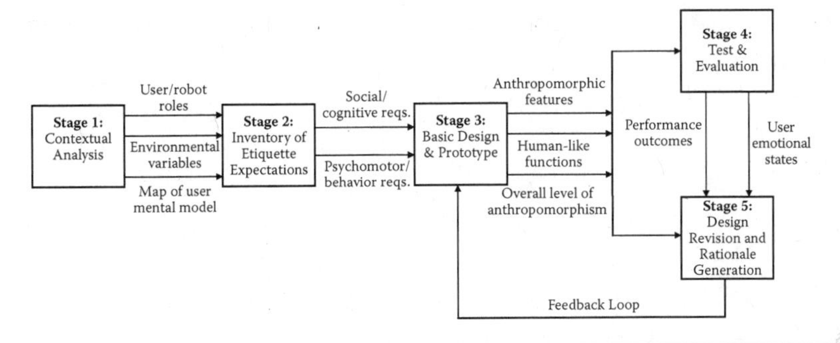

Figure 10.3 Five-stage design process for anthropomorphism and etiquette in HRI applications.

Stage 5—Feedback test and evaluation results into the design process: Finally, modifications to the robot interface design should be formulated to influence perceptions of anthropomorphism and etiquette in line with expectations. These modifications should be based on the performance outcomes and user emotional experiences recorded during the test process. Figure 10.3 presents a schematic diagram of the proposed design process and the information flows between stages.

10.5 Measures for Assessing Effectiveness of Etiquette in Robot Design

There is an increasing requirement for robots, working as assistants to people, to account for human emotional states and social needs. From a social interaction perspective, optimal HRI would involve robots naturally adapting to etiquette expectations of their human users (as humans do in their interaction with other humans). Traditional measures of HRI, like robot efficiency and effectiveness (Steinfeld et al., 2006), neglect human emotional responses and the fact that task performance is often motivated by emotions, environmental conditions, and cognitive factors (Picard and Klein, 2002). Emotional experiences activated during interaction with robots can be considered measures of robot adaptation to human etiquette expectations, as people naturally express emotions towards machines and often treat computer technology as social actors (e.g., Reeves and Nass, 1996). Just like in interaction between humans, appropriate social etiquette by a robot, such as establishing proper space from the user, maintaining eye contact, and providing supportive vocal or facial expressions, can lead to positive emotions for the human (or regulation of negative emotions). However, robots violating established rules or conventions of social etiquette (e.g., lack of attention to a speaker, lack of feedback in conversation, lack of pauses in vocalizations) may result in negative emotions for humans, such as frustration and anger. When etiquette issues arise in HRI (or HCI), performance outcomes may only indicate symptoms of the underlying problem. As the appropriate etiquette for a given context and human–robot collaboration is likely to be complex and mostly implicit (Miller, 2002), measures of user emotional responses combined with other measures implying consequences of robot etiquette may be necessary to determine the root cause/pathology of interaction breakdowns, including etiquette problems.

10.5.1 Role of Emotion and Etiquette in HRI

Human emotions can influence individual tasks requiring voluntary effort as well as human–robot collaboration. People experiencing negative emotions tend to focus on problematic details of interaction and may lose the broader "picture" on a collaboration. They also tend to stick to their initial interaction strategy and become increasingly tense and anxious when that strategy fails (Norman, 2004). Conversely, people with positive emotions are more tolerant of difficulties in interaction and more effective in searching for alternative approaches to collaboration, which may lead to satisfactory outcomes. Thus, robots causing positive emotional experiences for users may also be expected to support more productive collaborations with humans.

It has also been demonstrated that a computer can actively adapt to user frustrations it may have caused and extend the interaction so that the ultimate outcome may be more pleasurable and productive for the human (Klein, Moon, and Picard, 2002). Along this line, robot design should implement etiquette in an intelligent way in order to send the right signals to a human user at the right time to support overall system performance. However, it is important not to simply assume that a robot should be designed to behave exactly as a human would in similar situations (Walters, Dautenhahn, Woods, and Koay, 2007). Enhanced perception and recognition systems are being developed for robots (e.g., Breazeal, 2003; Breazeal and Aryananda, 2002), such that two-way interaction between humans and robots will be more prevalent in the future. Consequently, human rules of social conduct may be more rigorously applied to robots in the future. Etiquette in robot design, however, not only implies managing collaboration through expressions of emotion, but also negotiating and balancing initiative with human users, based on the robot's role (Horvitz, 1999; Ward and Marsden, 2004). This role may be different from roles typically assumed by humans in human–human interaction.

10.5.2 Models and Measures of Emotion for Evaluating HRI

In order to use human emotions in HRI as a vehicle for evaluating robot design, a model of emotion identifying specific states within an emotional space is needed. Three types of emotion models exist in the psychology literature, including: basic emotion theory (Ekman, 1992a; 1992b), dimensional emotion theory (Lang, 1995; Russell, 1980), and models from appraisal theory (e.g., Roseman, 2001). Basic emotion theory contends that there is a concise set of primary emotions (anger, disgust, fear, happiness, sadness, and surprise; see Peter and Herbon, 2006)) and that they are distinguishable from each other and other affective phenomena (Ekman, 1999). Ekman (1999) also proposed that each primary emotion represents a theme or a family of related emotion states from which other emotions can be constructed by variations.

Dimensional emotion theory argues that all emotions can be located in a two-dimensional space, as coordinates of affective *valence* and *arousal*. Russell (1980) described the typical course of an emotion, saying: (1) an event, or some combination of events (internal or external), produces a noticeable change in an individual's state; (2) the individual automatically perceives this change in his or her core affective state in terms of *pleasure–displeasure* and *arousal–sleep*; and (3) the individual seeks to label the emotional state within a dimensional space.

Appraisal theory emphasizes that emotions are elicited by evaluations (*appraisals*) of events and situations, and differentiated by a set of appraisal criteria (Lazarus, 2001; Scherer, 2001), including: novelty and expectancy, relevance and goal conduciveness (or motive consistency), agency or responsibility, and perceived control, power, or coping potential (van Reekum et al., 2004). This model and the others above relate different concepts of emotion and how emotions may be structured in theory. Consequently, they lead to different approaches of observing and assessing emotions in practice, based on either subjective or objective measures (Peter and Herbon, 2006).

10.5.3 Subjective Measures of Emotion

Subjective measures of emotion include self-reports, interviews on emotional experiences, and questionnaires on which a respondent identifies images of expressions or phrases that most closely resemble his or her current feelings. Specific subjective measures of, for example, arousal and valence include the Affect Grid (Russell, Weiss, and Mendelsohn, 1989) and the Self-Assessment Manikin (Bradley and Lang, 1994). Although such measures can allow for quick and seemingly discrete assessments of subject emotional experiences, both theoretical and methodological concerns have been raised with these types of measurements in applied emotion studies. First, discrete classification of emotions may not capture the continuous nature of emotion states. Second, in many human–human interaction (and HRI) situations mixed emotions are likely to be evoked rather than a single, pure emotion. Peter and Herbon (2006) said, in this case, the resulting categorization of emotion using subjective measures cannot be clear-cut and they questioned the approach of locating specific emotions in a two-dimensional space. They argued that study results based on verbal labeling of emotional states need to be treated carefully regarding their applicability as a basis for designing HCI systems. Third, individual differences in emotion self-awareness and experiences may lead to different interpretations of the same emotional stimulus. Finally, in HCI the stimuli for potential emotions usually vary less than in human–human interaction (e.g., even when critical errors occur in using an operating system, a simple dialog box is used to present a verbal warning), and the affective bandwidth in the interaction is limited—there are few channels for people to perceive and express emotions (Picard and Klein, 2002). Therefore, it is possible that users in an HCI (or HRI) context may not experience the same range of distinguishable emotions as they experience in their daily lives in interacting with other humans. Consequently, the use of subjective measures of emotion in HCI and HRI applications may not clearly reveal different emotional states.

10.5.4 Physiological Measures for Assessing Emotion

In addition to subjective measures, there is evidence suggesting that human physiology relates to emotional states or events in HCI (Frijda, 1986; Scheirer, Fernandez, Klein, and Picard, 2002). Physiological variables that have been found to be correlated with various emotions include: heart rate (HR), blood pressure, respiration, electrodermal activity (EDA) and galvanic skin response (GSR), as well as facial EMG (electromyography). These measures can complement subjective measures by identifying emotional features that subjects fail to note in verbal reports, either because they forgot or they considered the feature to be unimportant (Ward and Marsden, 2004). Although physiological measures can be noisy, less specific in human state identification, and difficult to interpret, there is ongoing research on emotional state identification based on physiological data.

Mandryk and Atkins (2007) presented a method for continuously classifying the emotional states of computer/video game users based on a set of physiological variables and the use of fuzzy logic, as well as an established psychological model, specifically the arousal–valence model. They developed a fuzzy logic model to transform HR, GSR, and facial EMG (smiling and frowning activity) into arousal and valence states. A second fuzzy logic model was then used to transform arousal and valence classifications into five lower-level emotional states related to the gaming situation, including: boredom, challenge, excitement, frustration, and fun. Results from the fuzzy logic models exhibited the same trends as self-reported emotions for fun, boredom, and excitement. The advantage of the fuzzy logic approach is that the emotions of a user can be quantitatively assessed from physiological data over an entire test trial, thus, revealing variance within a complete emotional experience.

As another example, van Reekum et al. (2004) used continuous psychological data in combination with appraisal theory to assess the experience of pleasantness and goal conduciveness of events in a computer game across different performance levels. Appraisal theory holds that appraisals lead to adaptive physiological changes that should be measurable using psychophysiological techniques. van Reekum et al. (2004) reported that skin conductance activity was higher for unpleasant sounds after event onset compared to pleasant sounds. It was also reported that conducive events were related to higher phasic skin conductance activity, more positive finger temperature slopes, and longer inter-beat intervals (i.e., lower heart rate), compared to obstructive events. The physiological variables, including inter-beat interval, pulse-transit time, and finger temperature slope, all showed a linear decrease with increasing performance level.

In summary, to capture a complete picture of user emotional experiences during interaction with computers or robots, objective (i.e., physiological) measures may need to be integrated with subjective measures to accurately assess specific emotional states and underlying etiquette issues (Champney and Stanney, 2007). Measures of emotion in HRI applications can and should be related to particular features and behaviors of the robot as a basis for directing improvements in system design for performance and etiquette.

10.6 Conclusions

There is a wide range of robot technologies currently available and being used for interactive and social applications, including those in health care and home environments. In the near future, service robot applications are expected to dramatically expand and become more complex due to increasing healthcare demands and the possibility of direct patient–robot interaction. The interface requirements of existing users, including nurses and support operators, include programming, monitoring, data collection, and navigation. Robot interaction with patients poses additional interface requirements to address patient emotional and social needs and the need for robots to conform with etiquette expectations.

With this in mind, the degree of anthropomorphism in robot design, and user perceptions of social etiquette, are considered to be critical aspects of design. Interface design guidelines exist for HRI applications but few address anthropomorphism needs and etiquette expectations of users. We proposed a simple design process to address these issues in interactive and social robotics. Research on anthropomorphism in design suggests that anthropomorphic features largely influence human perceptions of device behavior and social capability. Consequently, we believe anthropomorphism may drive the extent to which robots are perceived as having etiquette in social contexts as well as patient perceptions of the quality of performance and interaction.

Related to this, we identified human emotional experiences as an important measure of HRI effectiveness that has not been emphasized in prior research. Subjective and physiological response measures for inferring emotion states may be critical for assessing the quality of, for example, patient–robot interaction. Future research should make use of both physiological and subjective measures of emotion as an evaluation tool for interactive robot design in contexts including hospital nursing and home healthcare service. There is a need to determine the extent to which specific robot interface features support or compromise perceptions of etiquette and to formulate affective robot design guidelines.

References

Bethel, C. L. and Murphy, R. R. (2006). Affective expression in appearance-constrained robots. In *Proceedings of ACM SIGCHI/SIGART 2nd Conference on Human–Robot Interaction (HRI '06)* (pp. 327–328). New York: ACM Press.

Bradley, M. M. and Lang, P. J. (1994). Measuring emotion: The self-assessment manikin and the semantic differential. *Journal of Behavior Therapy and Experimental Psychiatry, 25*(1), 49–59.

Breazeal, C. (2002). *Designing Sociable Robots*. Cambridge, MA: MIT Press.

Breazeal, C. (2003). Emotion and sociable humanoid robots. *International Journal of Human-Computer Studies, 59*(1–2), 119–155.

Breazeal, C. (2003). Toward sociable robots. *Robotics and Autonomous Systems, 42*, 167–175.

Breazeal, C. and Aryananda, L. (2002). Recognition of affective communicative intent in robot-directed speech. *Autonomous Robots, 12*(1), 83–104.

Bruce, A., Nourbakhsh, I., and Simmons, R. (2002). The role of expressiveness and attention in human-robot interaction. In *Proceedings of the IEEE International Conference on Robotics and Automation (ICRA)* (pp. 4138–4142), Piscataway, NJ.

Burgoon, J. K., Bonito, J. A., Bengtsson, B., Cederberg, C., Lundeberg, M., and Allspach, L. (2000). Interactivity in human-computer interaction: A study of credibility, understanding, and influence. *Computers in Human Behavior, 16*(6), 553–574.

Champney, R. K. and Stanney, K. M. (2007). *Using emotions in usability*. In *Proceedings of the Human Factors and Ergonomics Society 51st Annual Meeting* (pp. 1044–1049), Santa Monica, CA.

Corliss, W. R. and Johnsen, E. G. (1968). Teleoperator Controls: An AEC-NASA Technology Survey (Tech. Report NASA-SP-5070). Washington, DC: NASA.

Dautenhahn, K., Woods, S., Kaouri, C., Walters, M. L., Koay, K. L., and Werry, I. (2005). What is a robot companion—friend, assistant or butler? In *Proceedings of IEEE IROS* (pp. 1488–1493), Edmonton, Canada.

Dennett, D. C. (1987). *The Intentional Stance*. Cambridge, MA: MIT Press.

DiSalvo, C. F. and Gemperle, F. (2003). From seduction to fulfillment: The use of anthropomorphic form in design. In *Proceedings of the International Conference on Designing Pleasurable Products and Interfaces,* Pittsburgh, PA. New York: ACM Press.

Duffy, B. R. (2003). Anthropomorphism and the social robot. *Robotics and Autonomous Systems, 42*(3–4), 177–190.

Ekman, P. (1992a). Are there basic emotions? *Psychological Review, 99*(3), 550–553.

Ekman, P. (1992b). An argument for basic emotions. *Cognition and Emotion, 6*(3–4), 169–200.

Ekman, P. (1999). Basic emotions. In T. Dalgleish and M. Power (Eds.), *Handbook of Cognition and Emotion*. Sussex, U.K.: John Wiley & Sons.

Engelberger, G. (1998). HelpMate, a service robot with experience. *Industrial Robot, 25*(2), 101–104.

Epley, N., Waytz, A., and Cacioppo, J. T. (2007). On seeing human: A three-factor theory of anthropomorphism. *Psychological Review, 114*(4), 864–886.

Evans, J. M. (1994). HelpMate: An autonomous mobile robot courier for hospitals. In *Proceedings of the International Conference on Intelligent Robots and Systems (IROS)* (pp. 1695–1700), Munich, Germany.

Foner, L. N. (1997). Entertaining agents: a sociological case study. In *Proceedings of the First International Conference on Autonomous Agents* (pp. 122–129). Marina del Rey, CA.

Fong, T., Nourbakhsh, I., and Dautenhahn, K. (2003). A survey of socially interactive robots. *Robotics and Autonomous Systems, 42*, 143–166.

Frijda, N. H. (1986). *The Emotions*: Cambridge University Press.

Goetz, J., Kiesler, S., and Powers, A. (2003). Matching robot appearance and behavior to tasks to improve human-robot cooperation. In *Proceedings of the 12th IEEE International Workshop on Robot and Human Interactive Communication (RO-MAN)* (pp. 55–60). Berkeley, CA.

Guthrie, S. E. (1997). Anthropomorphism: A definition and a theory. In R. W. Mitchell, N. S. Thompson, and H. L. Miles (Eds.), *Anthropomorphism, Anecdotes, and Animals* (pp. 50–58). Albany: State University of New York Press.

Hanson, D. (2006). Exploring the aesthetic range for humanoid robots. In *Proceedings of the ICCS/CogSci-2006 Symposium: Toward Social Mechanisms of Android Science* (pp. 16–20), Vancouver, Canada: Cognitive Science Society.

Hinds, P. J., Roberts, T. L., and Jones, H. (2004). Whose job is it anyway? a study of human-robot interaction in a collaborative task. *Human-Computer Interaction, 19*(1&2), 151–181.

Horvitz, E. (1999). Principles of mixed-initiative user interfaces. In *Proceedings of the ACM SIGCHI Conference on Human Factors in Computing Systems (CHI '99)* (pp. 159–166). New York: ACM Press.

Jansen, J. F. and Kress, R. L. (1996). Hydraulically powered dissimilar teleoperated system controller design. In *Proceedings of the IEEE International Conference on Robotics and Automation (ICRA)* (pp. 2484–2491). Piscataway, NJ: IEEE.

Johnsen, E. G., and Corliss, W. R. (1971). *Human Factors Applications in Teleoperator Design and Operation*. New York: Wiley-Interscience.

Kanda, T., Kamasima, M., Imai, M., Ono, T., Sakamoto, D., Ishiguro, H. et al. (2007). A humanoid robot that pretends to listen to route guidance from a human. *Autonomous Robots, 22*(1), 87–100.

Khan, Z. (1998). Attitudes towards Intelligent Service Robots. Technical Report No. TRITA-NA-P9821, NADA, KTH, Stockholm, Sweden.

Kiesler, S. and Goetz, J. (2002). Mental models of robotic assistants. In *Proceedings of the ACM SIGCHI Conference on Human Factors in Computing Systems (CHI '02)* (pp. 576–577). New York: ACM Press.

Klein, J., Moon, Y., and Picard, R. W. (2002). This computer responds to user frustration: Theory, design, and results. *Interacting with Computers, 14*(2), 119–140.

Kochan, A. (1997). HelpMate to ease hospital delivery and collection tasks, and assist with security. *Industrial Robot, 24*(3), 226–228.

Krishnamurthy, B. and Evans, J. (1992). HelpMate: A robotic courier for hospital use. In *Proceedings of the IEEE International Conference on Systems, Man and Cybernetics* (pp. 1630–1634). Piscataway, NJ: IEEE.

Lang, P. J. (1995). The emotion probe: Studies of motivation and attention. *American Psychologist, 50*(5), 372–385.

Langer, E. J. (1992). Matters of mind: Mindfulness/mindlessness in perspective. *Consciousness and Cognition, 1*(4), 289–305.

Lazarus, R. S. (2001). Relational meaning and discrete emotions. In K. R. Scherer, A. Schorr, and T. Johnstone (Eds.), *Appraisal Process in Emotion: Theory, Methods, Research*. New York, NY: Oxford University Press.

Luczak, H., Roetting, M., and Schmidt, L. (2003). Let's talk: Anthropomorphization as means to cope with stress of interacting with technical devices. *Ergonomics, 46*(13), 1361–1374.

Mandryk, R. L. and Atkins, M. S. (2007). A fuzzy physiological approach for continuously modeling emotion during interaction with play technologies. *International Journal of Human-Computer Studies, 65*, 329–347.

Miller, C. A. (2002). Definitions and dimensions of etiquette. In *Proceedings of the AAAI Fall Symposium on Etiquette for Human-Computer Work* (pp. FS-02-02) Menlo Park, CA: American Association for Artificial Intelligence.

Mori, M. (1982). *The Buddha in the Robot.* Charles E. Tuttle.

Nass, C. and Moon, Y. (2000). Machines and mindlessness: Social responses to computers. *Journal of Social Issues, 56*(1), 81–103.

Norman, D. A. (1999). Affordances, conventions and design. *Interactions, 6*(3), 38–43.

Norman, D. A. (2004). *Emotional Design: Why We Love (or Hate) Everyday Things.* New York: Basic Books.

Novak, J. D. and Gowin, D. B. (1984). *Learning How to Learn.* New York: Cambridge University Press.

Peter, C. and Herbon, A. (2006). Emotion representation and physiology assignments in digital systems. *Interacting with Computers, 18,* 139–170.

Picard, R. W. and Klein, J. (2002). Computers that recognise and respond to user emotion: Theoretical and practical implications. *Interacting with Computers, 14*(2), 141–169.

Pollack, M. E., Brown, L., Colbry, D., Orosz, C., Peintner, B., Ramakrishnan, S. et al. (2002). Pearl: A mobile robotic assistant for the elderly. In *Proceedings of the AAAI Workshop on Automation as Eldercare* (pp. WS-02-02) Menlo Park, CA: American Association for Artificial Intelligence.

Reeves, B. and Nass, C. (1996). *The Media Equation: How People Treat Computers, Television, and New Media Like Real People and Places.* New York: Cambridge University Press.

Roseman, I. J. (2001). A model of appraisal in the emotion system. In K. R. Scherer, A. Schorr, and T. Johnstone (Eds.), *Appraisal Process in Emotion: Theory, Methods, Research.* New York, NY: Oxford University Press.

Russell, J. A. (1980). A circumplex model of affect. *Journal of Personality and Social Psychology, 39*(6), 1161–1178.

Russell, J. A., Weiss, A., and Mendelsohn, G. A. (1989). Affect grid: A single-item scale of pleasure and arousal. *Journal of Personality and Social Psychology, 57*(3), 493–502.

Schaeffer, C., and May, T. (1999). Care-O-bot: A system for assisting elderly or disabled persons in home environments. In *Proceedings of AAATE-99* (pp. 340–345). Amsterdam, the Netherlands: IOS Press.

Scheirer, J., Fernandez, R., Klein, J., and Picard, R. W. (2002). Frustrating the user on purpose: A step toward building an affective computer. *Interacting with Computers, 14*(2), 93–118.

Scherer, K. R. (2001). Appraisal considered as a process of multilevel sequential checking. In K. R. Scherer, A. Schorr, and T. Johnstone (Eds.), *Appraisal Processes in Emotion: Theory, Methods, Research* (pp. 92–120). New York: Oxford University Press.

Sheridan, T. B. (1989). Telerobotics. *Automatica, 25,* 487–507.

Shneiderman, B. (1989). A nonanthropomorphic style guide: Overcoming the humpty-dumpty syndrome. *The Computing Teacher, 16*(7).

Steinfeld, A., Fong, T., Kaber, D. B., Lewis, M., Scholtz, J., Schultz, A. et al. (2006). Common metrics for human-robot interaction. In *Proceedings of the 2006 ACM Conference on Human-Robot Interaction (HRI '06).* New York: ACM Press.

Turkle, S. (1984). *The Second Self: Computers and the Human Spirit*. New York: Simon and Schuster.

van Reekum, C. M., Johnstone, T., Banse, R., Etter, A., Wehrle, T., and Scherer, K. R. (2004). Psychophysiological responses to appraisal dimensions in a computer game. *Cognition and Emotion, 18*(5), 663–688.

Walters, M. L., Dautenhahn, K., Woods, S. N., and Koay, K. L. (2007). Robotic etiquette: Results from user studies involving a fetch and carry task. In *Proceedings of the ACM/IEEE International Conference on Human-Robot interaction (HRI '07)* (pp. 317–324). New York, NY: ACM Press.

Walters, M. L., Syrdal, D. S., Dautenhahn, K., te Boekhorst, R., and Koay, K. L. (2008). Avoiding the uncanny valley: Robot appearance, personality and consistency of behavior in an attention-seeking home scenario for a robot companion. *Autonomous Robots, 24*(2), 159–178.

Ward, R. D. and Marsden, P. H. (2004). Affective computing: Problems, reactions and intentions. *Interacting with Computers, 16*(4), 707–713.

Woods, S., Dautenhahn, K., Kaouri, C., Boekhorst, R. T., Koay, K. L., and Walters, M. L. (2007). Are robots like people?—Relationships between participant and robot personality traits in human–robot interaction studies. *Interaction Studies, 8*(2), 281–305.

PART V

Understanding Humans: Physiological and Neurological Indicators

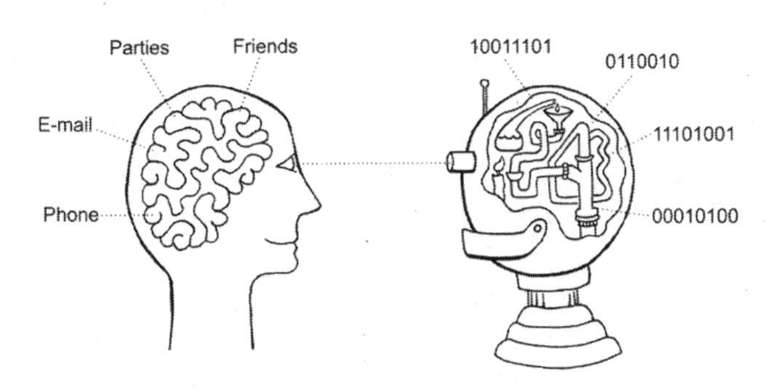

11

The Social Brain

Behavioral, Computational, and Neuroergonomic Perspectives

EWART DE VISSER AND
RAJA PARASURAMAN

Contents

11.1 Introduction

The goal of this chapter is to review recent neurophysiological findings and assess their relevance for human–automation etiquette. We will explore different facets of politeness and etiquette and how these influence collaboration between humans and automation. We will identify theories relevant to understanding the nature of human–automation etiquette and explore several ways to evaluate how etiquette affects performance in complex human–machine systems, including behavioral and computational methods. We complement these approaches with a *neuroergonomic* perspective (Parasuraman, 2003; Parasuraman and Rizzo, 2007). Recent research has shown that examining theories and empirical findings on human brain function can bolster our understanding of such aspects of human performance as mental workload, vigilance, adaptive automation, and individual differences (Parasuraman and Wilson, 2008). Accordingly, we examine human–automation etiquette from the perspective of both human behavior and neural mechanisms. Finally, we suggest several applications of research on human–automation etiquette.

11.2 Background

Autopilots in the cockpit and controlled emergency shutdowns in nuclear power plants represent some of the earliest forms of industrial automation, dating to the 1950s and 1960s. These automated systems, while capable of more reliable, timely, and precise control than humans, possessed few of the features commonly ascribed to human intelligence. As such, even though it is now commonplace to speak of human "interaction" with automation, human usage of most automated systems typically involved no more social dynamics or interpersonal exchange than would occur in using a tool such as a hammer or a laptop computer. Yet, when a blunder or fault occurs—hitting one's thumb with the hammer or the laptop freezing up at an inconvenient time—human users may blame themselves in the former case but the machine in the latter (Miller, 2004)—as if the computer were a bumbling human subordinate. The distinction seems to be related to the human tendency to ascribe agent-like qualities (such as awareness, volition, and intent) to tools such as computers (but not hammers) that are capable of a degree of autonomous behavior.

Nass and colleagues have also shown that the way people behave towards computers shows striking similarities to their interaction with other people. When Nass et al. (1994) examined how people responded to a computer rating its own performance on a task, participants rated the computer as more competent and friendly when they filled out the ratings on the same computer they had used for the test. Ratings were significantly lower when participants responded with pencil and paper or on another computer. The authors proposed that this finding generalizes the social norm politeness principle to computers. Remarkably, participants did not think they would rate the computer higher when they were told about the experimental manipulation. The authors concluded that computers are often seen as social actors if they behave similarly to humans.

Recent technological advances have led to computer systems possessing even greater agent-like qualities than the simple computers used in the studies by Nass and colleagues. Whereas at one time automation primarily involved sensing and mechanical actions, modern automation often entails information-processing or decision-making functions. The trend began in aviation but has extended to most domains at work, the home, transportation, and many forms of entertainment. Furthermore, older automation, while able to sense the world and control action, was immobile. But modern robots and unmanned vehicles possess eyes, ears, hands, feet, as well as rudimentary thinking brains. Robots such as Asimo (Honda, 2007) and other animated agents provide a compelling example of how agent-like qualities encourage humans to interact with computers in human-like ways. Asimo can speak and can be spoken to. It can recognize natural movements of people such as handshakes and walking, identify sounds in the environment, and remember faces. This allows Asimo to bring coffee to co-office workers, greet people when they walk by, and avoid collisions by yielding to a person. Asimo and other "service robots" are being developed to assist humans in a variety of ways. Etiquette effects are not limited to human–robot interaction though but extend to nonembodied agents such as animated agents and traditional decision aiding automation (Parasuraman and Miller, 2004).

Many of the cognitive processes underlying etiquette have a basis in neurophysiology. Concepts related to etiquette that have a firm basis in neuroscience include trust, emotions, and arousal. It is likely that exploration of data in this literature will provide informative results for the human–automation interaction field of study for several reasons. If the goal of "design for human–machine etiquette" is to improve machine design in its etiquette toward humans, one way is to improve etiquette measures to help machines assess how effective they are in their "manners." Measures based on neurophysiological data and computational models may help achieve this goal. Furthermore, neuroscience techniques can help assess etiquette effects for human–human or human–machine interaction.

11.3 A Theoretical Framework for Politeness and Etiquette

Brown and Levinson (1987) proposed a comprehensive theory of politeness in human relations. They postulated that the act of politeness is a rational and necessary act, and that all social actors attempt to protect their face, defined by them as "the public self-image that every member wants to claim" for him or herself. Many actions that are expressed from one social actor to another can be considered a face-threatening act or FTA. For example, if Asimo asks an office worker to "move out of the way" its expression can be classified as an FTA towards the office worker's negative face; Asimo's request impedes the office worker's ability to move as she pleases. If Asimo said "You should be working right now," this expression can be classified as an FTA towards the office worker's positive face; Asimo disapproves of her behavior. Thus, FTAs differ in their severity. If for example, Asimo were to order the office worker to "Stop your work," that would constitute a much more severe FTA on her negative face than if it were to ask her to remember her next appointment. The goal of etiquette use is to minimize the impact of the FTAs. Whether this attempt is potentially effective when computationally modeled is an issue we turn to next.

11.3.1 A Computational Model of Politeness

Brown and Levinson partially instantiated their theory in a computational model that allows for calculating the impact of an FTA. Miller

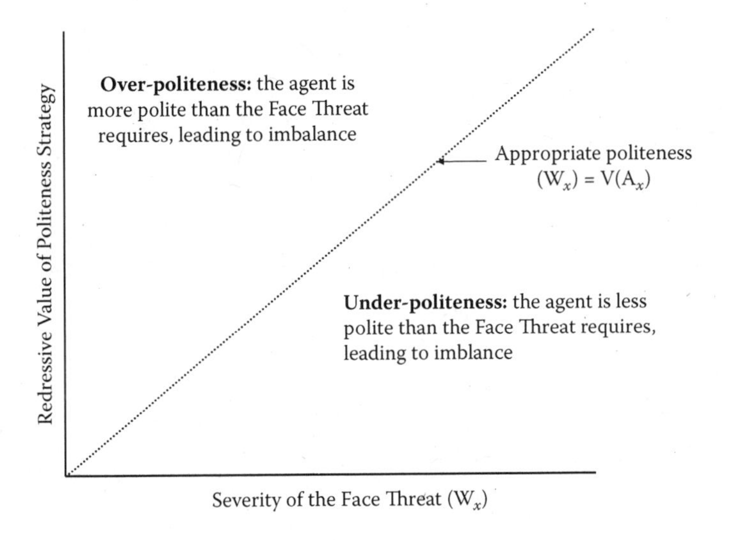

Figure 11.1 Calibration of politeness strategy to the severity of face threat.

et al. extended this approach and provided a more comprehensive and formal computational model of politeness. The goal of this model is to show how FTAs may be mitigated by politeness strategies. This may be expressed mathematically as follows:

$$W_x \cong V(\mathbf{A}_x) \tag{11.1}$$

where W_x is the severity of the FTA threat. Factors influencing the FTA severity include social distance, the power that the speaker exerts over the listener, and the absolute ranking of imposition in a given culture. $V(\mathbf{A}_x)$ is the combined redressive value of the politeness strategies $(\mathbf{A}_x)$.

In an ideal and balanced world, the severity of the FTA threat is more or less matched with an appropriately strong politeness strategy. Consequently, from the formula we can derive that if $W_x >> V(\mathbf{A}_x)$, the FTA has been under-addressed and the behavioral expression may be perceived as inappropriate, insufficient, or even rude. If, however, $W_x << V(\mathbf{A}_x)$, the FTA may be over-addressed and the expression in this situation may be perceived as awkward, or overly polite (see Figure 11.1).

11.3.2 Validating the Computational Model

The challenge in actually implementing the model described above is in measuring the values of the FTA threat and the effectiveness of the accompanying redressive action. Most elements in the formula are subjective and depend on an individual's perception of the situation or the *perceived etiquette.* A solution is to have expert raters, familiar with the Brown and Levinson model, evaluate certain utterances objectively and then verify if naïve raters rate them similarly, a procedure that was followed by Miller et al. (2007). While this is a rough, nonreal time prediction of perceived etiquette, high interobserver reliability for several situations has been observed between expert and naïve raters, providing initial evidence that the model can predict the outcome of several situations (Miller et al., 2007).

This method, however, only addresses consistency with raters who are not part of the actual vignettes. We may expect different ratings when participants are actually involved in the situation themselves and *feel* the threat and the ensuing politeness. Additionally, the model does not address how face threats rank in importance compared to the other demands of a given situation, for example, when people are engaged in life-critical tasks such as medical emergencies or military combat situations such as those described in Chapter 7 by Dzindolet et al. and Chapter 12 by Dorneich et al. Such an expansion is critical for understanding situations in which humans and automation are accomplishing tasks together. And finally, it may be difficult to obtain accurate weights for each of the components of the formula.

Nonetheless, the model serves as a good starting point for evaluating politeness in more naturalistic situations. The actual weighting of both the strength of the FTA and the subsequent redressive action may have to be further formalized. Another, perhaps promising, technique would be to evaluate the imbalance resulting from an inequality between severity of FTA and redressive actions. If, for example, the observer perceives a large imbalance between the FTA and the redressive action, we may observe behavioral and physiological markers that reflect this perceived imbalance. Such markers may include differences in trust, emotion, and arousal levels within the observer. In the next sections, we propose how neuroergonomic measures may provide a useful mechanism enabling computer agents to accurately assess how their actions have actually been perceived, so they may adjust their behavior accordingly.

11.4 Neuroergonomic Perspectives on Human–Automation Etiquette

While perceived etiquette per se has not been examined using cognitive neuroscience tools, the neural basis of human social interaction, as an intrinsic part of the emerging field of social neuroscience, has gained much recent attention. Information about how the human brain represents decisions in social contexts may inform the nature of decision making with regard to human–machine interactions. This is particularly relevant because violations of perceived etiquette may lead to a recalibration of trust and result in a change in human–automation interaction (Parasuraman and Miller, 2004).

A major issue in social neuroscience is how the human brain determines the intent of others; determining intent accurately enables a person to act appropriately with others in social situations. Human users of computer software frequently attribute intentionality to computers (Nass, 2004); the reverse, enabling computers to understand human intent (or at least reactions) is likely a necessary requirement for allowing computers to respond appropriately to humans. Determining intent is thought to be made possible by the human capacity to attribute mental states to oneself and to others, an ability that has been called *theory of mind* (Baron-Cohen, 2000). Theory of mind underlies our ability to understand the behavior of others based on how they are feeling, what they intend to do, and how they are likely to act (Adolphs, 2003).

Reading intentions in others is cued through the use of social signals (such as changes in facial expression, posture, or tone of voice; see Chapter 9 by Bickmore). An accurate perception of these social signals is integral to the development of social relationships, and a fundamental requirement for appropriate etiquette strategies. The recognition of emotion in others and our own emotional reactions drive the majority of our daily social interactions. However, while people recognize these social signals easily and naturally, they may be difficult and computationally expensive for computer agents to recognize through visual and auditory processing methods. An alternate approach may be to sense the physiological and neurological signals associated with social relationships and interactions. Recently, much effort has been devoted to the identification of the neural mechanisms underlying the perception of social signals, particularly as related to the perception of emotion. Thus, if inappropriate application of etiquette leads to strong emotional responses in people, then computer agents may be able to use measurement of emotional reactions as a way to assess when they have inadvertently committed an etiquette violation.

We will explore this hypothesis by reviewing findings in the area of etiquette and social neuroscience, and discussing theoretical and neural mechanisms of trust. We also discuss the processing of emotion, as well as of social decision making, and identify some of the controlling brain regions: the amygdala and the orbitofrontal cortex.

11.4.1 Etiquette and Social Neuroscience

In our everyday social interactions we utilize a variety of verbal (language) and nonverbal social signals, ranging across modalities and categories. These include facial expressions, body posture, direction of gaze, biological motion, touch, voice, prosody, etc. A wealth of information has been discovered about each of these signals. However, two of them, facial expression and prosody (the rhythm, stress, and intonation in speech) are particularly relevant to etiquette, and constitute a significant portion of the current knowledge of emotional processing. fMRI research in face perception has systematically reported the involvement of the fusiform gyrus association with the encoding of the structural, static properties of faces, while the superior temporal gyrus (STG) and superior temporal sulcus (STS) seem to modulate the processing of the dynamic configurations of faces such as facial expressions, and eye and mouth movements (McCarthy, 1999; Haxby et al., 2000).

Complementary insights come from the timing of face processing as identified in studies using event-related potentials (ERPs) and magnetoencephalography (MEG). These show that the earliest activity that seems to discriminate between emotional facial expressions is seen in the midline occipital cortex as early as 80 ms to 110 ms (Pizzagalli et al., 1999; Halgren et al., 2000). Moreover, these responses may be modulated by feedback to the visual cortex from the amygdala and orbitofrontal cortex. The processing of emotional prosody, the intonation of speech, or the vocal expression of emotion has been identified as recruiting many of the regions also correlated with the processing of fearful facial expressions, such as the amygdala.

Integrative findings from the perception of social signals may be useful in the design of personified machines that utilize these signals in order to establish rapport with their human operators. For example, Asimo could register a certain emotional expression and use this information to form a politeness strategy or an individual emotional profile of the person in question. Other aspects of etiquette such as politeness and interruptiveness that are more characteristic of high criticality automation systems are still novel open ground for investigation.

One goal for developing such systems is to promote human understanding of the machine and potentially develop "real" relationships with automation similarly to how humans form relationships with one another. A positive relationship between humans and automation will lead to better understanding between the two and better future design. In particular, trust between humans and machine is critical to developing such a relationship, much like trust is important in developing relationships between humans. In the next section we focus our discussion on recent research on the topic of trust and suggest ways this may be useful to the research in the design of machines that can exhibit appropriate etiquette.

11.4.2 Neural Mechanisms of Trust

Trust has been an important factor in understanding human-automation interaction—even before automation possessed a face, could move, or exhibit a "personality." Two factors were important in motivating interest in studies of human trust of automation. First, despite the best intentions of designers, automated systems have typically not been used in ways that they were intended, with instances of misuse and disuse being prevalent. Second, because most automated systems do not exhibit 100% reliability, variable human trust is the consequence of imperfect automation. In this section, we review some of the recent advances in the understanding of the neural basis of social and reward learning and expectation processes associated with trust. Moreover, the potential application of the insights arising from current and future findings of this social issue to the design of human–machine interfaces will be discussed.

Various methodologies and conceptual approaches have been used to examine the neural mechanisms of trust. We will discuss a few of these studies which we believe illustrate significant advances in three important dimensions: (1) a theoretical framework, (2) hormonal orchestration, and (3) cerebral neural networks associated with the processing of this psychological construct.

11.4.2.1 A Theoretical Framework Trust significantly impacts the effectiveness of collaboration between people and computers, particularly in willingness on the part of humans to share or delegate tasks to computers, and create an impetus for supportive behavior. For the purposes of this chapter we will adopt the definition of trust given by Lee and See (2004): "the attitude that an agent will help achieve an individual's goals in a situation characterized by uncertainty and vulnerability." This definition fits the idea that social actors are rational and that individuals maintain some sense of face (i.e., vulnerability).

Empirical evidence showing the impact of polite strategies on trust has been demonstrated in recent studies (Parasuraman and Miller, 2004; Miller et al., 2007; Wu et al., Chapter 4 in this volume). In Parasuraman and Miller (2004), participants were required to perform a multiple task battery simulating different flight tasks. One of the tasks involved diagnosing and correcting engine malfunctions. The other two tasks involved tracking and fuel management. The engine monitoring task was supported by an automated fault diagnosis system that provided advisories concerning different engine parameters and possible reasons for the malfunction. When providing advice, the automation either displayed good etiquette (defined as being patient and noninterruptive) or poor etiquette (defined as being impatient and interruptive). In addition, the advisories provided by the automation were either correct most of the time (high reliability) or only some of the time (low reliability).

When participants interacted with this automation, their trust was lower in the poor etiquette condition compared to the good etiquette condition. Interestingly, in the low reliability condition, good etiquette increased participants' rating of trust close to the trust levels in the highly reliable, but poor etiquette condition (see Figure 11.2). An

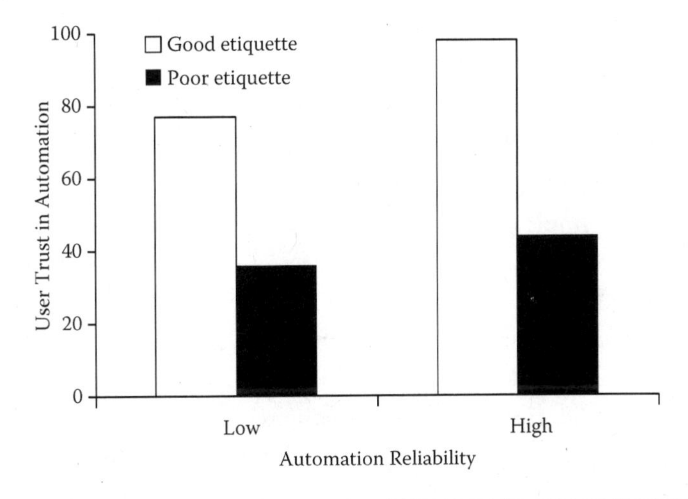

Figure 11.2 Effects of automation etiquette and automation reliability on subjective reports of trust in automation.

almost identical effect for trust was found in a recent etiquette study using simulated unmanned aerial vehicles (de Visser et al., 2009). This suggests that when the automation is designed to exhibit good etiquette, it has a positive impact on collaboration behaviors. Important to note here is that interruption per se, which can be annoying, could not explain this effect. Participants experiencing neutral interruptions had lower trust compared to the good etiquette conditions.

As shown above by previous studies, etiquette and trust have a close relationship. Trust will be engendered when appropriate etiquette is used and will decrease or disappear when inappropriate etiquette is used.

But why is trust impacted by etiquette behaviors? Many trust definitions include an element of vulnerability on the part of the trustor. Social actors evaluate etiquette behavior to assess this vulnerability by observing another agent's reliability and dependability. Etiquette behavior can be evaluated in two distinct ways. Displaying appropriate etiquette appeals at an emotional level because appropriate etiquette can increase positive feelings; inappropriate etiquette can increase negative feelings. Social agents can also evaluate etiquette behavior analytically by considering the display of etiquette as evidence of regard from the observed and perhaps even of the actor's expertise for rules of propriety. This analytical process is akin to *swift trust*, which can develop between a team of experts who can quickly establish whether team members possess a certain expertise on which they can rely.

11.4.2.2 Hormonal Orchestration Trust formation and attribution demands memory for prior interactions with an agent and ability to recognize that agent in the future. Thus, social memory (i.e., person/social recognition) is fundamental for adequate performance of cooperative behaviors and ultimate survival. Crucial stored information ranges from simple recognition of your kin, friends, and enemies to the specific complex social interactions and events experienced by any given individual. Social memory therefore functions as the basis of adaptive behaviors such as danger avoidance, cooperation, and reciprocal altruism and is pivotal for behavioral economic game theories (Axelrod and Hamilton, 1981). Furthermore, it has been proposed that social environmental factors, such as stress, can affect social memory performance. Takahashi (2005) reported seminal findings on this topic, linking behavior and physiology (Takahashi et al., 2004) and adding another valuable factor to this interaction: trust. More specifically, the correlation between hormonal orchestration (cortisol levels) and social memory performance was investigated evaluating the effects of interpersonal trust. A positive correlation was reported between social stress, high cortisol levels, and impaired social memory and a negative correlation between interpersonal trust and an increase of cortisol levels. On this basis, the authors proposed that interpersonal trust plays a significant role in social cooperation due to its effects on this hormonal orchestration mechanism, which lowers social stress, prevents the elevation of the cortisol levels, and thus ultimately allows for better social memory performance.

Are these peripheral hormonal changes reflected in or controlled by neural activity? Studies of oxytocin have provided clues in this regard. Oxytocin is a small peptide that acts as a neurotransmitter, with a dual role as a hormone and a neuromodulator. It is produced in the hypothalamus and targets a variety of brain areas associated with both emotional and social aspects of human behavior (i.e., amygdala, nucleus accumbens, etc.). Its presence and influence has also been studied in animals, where oxytocin has been found to influence male–female and mother–infant bonding and to modulate approach behavior. By diminishing the natural reluctance to avoid the close approach of another individual, thereby increasing trust in that person, oxytocin also facilitates the development of social cohesion and cooperation (Carter, 1998; Insel and Shapiro, 1992).

In a recent study, Kosfeld and colleagues (2005) explored the interaction between individuals playing the roles of either "investor" or "trustee" during a monetary exchange. Their results show that investors prepared with oxytocin show higher trust levels in the trustees than a placebo group. On the other hand the behavior of the trustees is not altered by oxytocin. Based on these results, it was concluded that this neurotransmitter can elicit an increase in trust without affecting the objective reliability of the agent. It was hypothesized that during real human interactions oxytocin is released in specific brain regions when a person perceives certain social configurations. This in turn leads to modulation of neural networks to produce an increase of trusting behavior related to another individual or group. This mechanism would act as a facilitator for interpersonal interactions that may contribute to both the individual and group's success.

Although we argue that insights from the neural basis for social interactions, such as trust, can provide significant insights into how to better develop human–machine interfaces, it is crucial to carefully test, and note, which components and vectors of these interactions can be reliably translated from human–human to a human–machine interaction. For instance, in this particular case, the role of oxytocin seems to be "social specific," and it may not represent a broader mechanism for risk sensitivity reduction In a controlled experiment in which the investors knew that the "trustee" was in fact a computer, no effect was observed. This result would seem to contradict the behavioral findings of Nass et al. who claims that general social norms are frequently applied to computers.

11.4.2.3 Neural Networks—The Caudate Nucleus As previously stated, the expression, prediction, and reciprocity of trust are paramount factors in social interactions related to cooperative behaviors (Axelrod and Hamilton, 1981; Coleman, 1990). As part of extremely complex social interactions it becomes remarkably difficult to design experimental paradigms that are able to rigorously assess this parameter, separating its effects from other factors. The most successful attempts, so far, in paradigm design have been based on economic exchange games (Berg et al., 1995; Camerer and Weigelt, 1988; Fehr et al., 1993).

Multiple studies have been carried out using similar trust/economic exchange game paradigms to explore the brain neural networks associated with reward prediction and trust (Adolphs, 2003). The reported neural networks include several brain areas. Specifically relevant for the subject of this chapter is the neural processing associated with the dorsal and ventral striatum and the anterior and middle cingulate cortices.

Behavioral studies have consistently shown what one would intuitively assume: if one player expresses reciprocity towards the other, his/her partner will be more likely to express trust in the future. More recently, fMRI studies have uncovered the neural basis of this behavior (King-Casas et al., 2005). A structure in the dorsal striatum, the caudate nucleus, which has been linked to learning and memory in both humans and animals (Packard and Knowlton, 2002; Poldrack, 2001; Shohamy, 2004; White, 1997), shows activity correlated with changes in reciprocity. More specifically, the fMRI BOLD response in the caudate nucleus was higher for benevolent than for malevolent reciprocity, suggesting that this brain structure computes information associated with fairness of a partner's decision.

Furthermore, when a player has positive experiences throughout the game, this signal shifts in its timing. It may precede the revelation of the investment. Taken together these findings suggest that the caudate activity could predict an *intention to trust* (King-Casas et al., 2005), complementing other recent brain imaging findings that associate activity in the caudate and putamen with reward prediction signals (McClure et al., 2003; O'Doherty et al., 2004; Pagnoni et al., 2002; Seymour et al., 2004). These studies support the possibility that specific patterns of brain activity—for example signals from the caudate nucleus—might be used as a fairly reliable predictor of the intention to trust an interactive partner. However, again one must apply extreme caution when depending on these correlations, since they can be extremely fragile and strongly influenced by a variety of other internal and external factors.

Delgado and colleagues (2005), also investigated the role of the striatum in adjusting future behavior on the basis of a reward feedback. They demonstrated that previous moral and social information about an individual can affect the effective processing of the neural circuitry of his/her interacting partner, in some cases annulling these predictive correlations. In this study, participants were given previous information about their hypothetical partners, depicting them as "good," "bad," or "neutral," and their choices during a trust game were recorded, both behaviorally and neurophysiologically (fMRI). As expected, participants were more cooperative, displaying more trust, with the morally "good" partners and taking less chances with the "bad" ones. Moreover, even when tested with equivalent reinforcement rates for all three moral categories, participants would persistently display more trust in the previously depicted "good" partners.

Interestingly, the caudate nucleus showed robust differential activity between positive and negative feedback, consistent with the hypothesis that it is involved in learning and adaptive choices (O'Doherty et al., 2004; Tricomi et al., 2004) but only for the "neutral" partner. There was no significant differential activity for the "good" partner, and it was very weak for the "bad" partner. This result shows that when an individual has an *a priori* moral expectation of his/her interacting partner, this belief not only influences the behavioral choices but also disrupts the regular encoding of evidence, diminishing reliance on feedback mechanisms and their associated neural circuitry (Delgado et al., 2005). It seems that the existence of *moral information* prior to an interaction reduces the differential activity in the caudate nucleus. This leads the individual to discount feedback information, and fail to adapt choices.

This is a striking example of the flexibility and potential unreliability of these systems as "strong" predictors.

Taken together, this research has several implications for measuring and studying perceived etiquette. First, if a machine is initially polite to a human then he or she is likely to reciprocate. Second, in order to promote trust, the machine must act in a fair manner according to a set of transparent rules. Third, it may be possible to use measurements of activation in the caudate nucleus to predict a human's intention to trust the machine (if indeed this type of trust is exhibited towards machines). Additionally, the human's trust in the machine may be influenced by the machine's appropriate or inappropriate use of etiquette. Finally, prior information about a person (or machine) may influence users' perceptions of the machine. When such information is available, politeness may not help to improve trust because the prior information may weigh more than any newly performed behavior by the agent.

11.4.3 Neural Systems Involved in the Processing of Emotion

In this section, we will discuss emotions, the parts of the brain and body that process emotions, and the way in which emotions exhibit themselves as measurable indicators of arousal in the brain and body, for example galvanic skin response (sweating). It may be possible to use such indicators to measure emotional responses; however, the connection between emotions and measurable responses is a complex one that is still incompletely understood.

11.4.3.1 Emotional Valence Information that is collected to form a trust judgment can be evaluated on the basis of the associated emotional valence. *Valence* refers to the direction of an emotion, which can be either positive or negative. Emotions can serve as quick indicators of the perceived value and purpose of certain information. In the context of etiquette, they can be used as a measure to evaluate whether a politeness strategy appropriately addressed the FTA. There has been much debate in the history of psychology about the number and type of emotions that humans experience. However, a useful and practical distinction is to classify emotion by either positive or negative affect (but see Lerner and Keltner, 2000). Although it may seem that these two aspects of emotion are opposites on a continuum, research suggests that they arise from different systems in the brain.

One well-studied construct closely related to negative affect is stress. Studies of stress have shown that people tend to cope with stressors by attempting to restructure tasks to avoid extremes of overload or underload. This type of adaptive response is generally but not always successful (Hancock and Warm, 1989). Performance effects of stress are well documented (Hancock and Szalma, 2008) and include narrowing of attention and working memory performance loss, but moderate stress may sometimes help in focusing attention on a particular aspect of the task.

The impact of positive affect on human performance has also received ample attention. In general, the effects of positive affect include attentional broadening, creative problem solving, cognitive flexibility, and situational coping skills. One proposed theoretical explanation for the overall beneficial effects of positive affect is the release of the neurotransmitter dopamine throughout the brain in such areas as the hippocampus, the amygdala, the striatum, the anterior cingulate cortex, and the prefrontal cortex. These areas have also been proposed by others to play an important role in emotional processing, in particular as identified in the work by Damasio and colleagues. These theories are relevant to assessment of human reactions to computer etiquette in human–computer interactions. Excessive politeness or lack of sufficient politeness may produce negative arousal, distrust, or stress. Appropriate politeness, where the level of the redress is in keeping with the degree of the face threat perceived, may give rise to positive affect. In the following sections we will discuss how such emotions are processed at a neural level in more detail.

11.4.3.2 The Amygdala and Orbitofrontal Cortex While many brain regions such as the anterior cingulate gyrus, hypothalamus, basal ganglia, insular cortex, and somatosensory cortex are involved in the processing of emotional stimuli and tasks, the orbitofrontal cortex and the amygdala have been identified as brain regions whose primary functions are related to the processing of emotion (Gazzaniga, Ivry, and Mangun, 2002).

The amygdala is a small structure in the medial temporal lobe adjacent to the anterior portion of the hippocampus, and has been primarily associated with emotional learning and memory processes. This modulation seems to be obtained by the coupling of emotional valence (affiliative or agonistic)/social value (positive or negative) to objects (Dolan, 2002). Furthermore, the amygdala's role in the neural modulation of affect seems to transcend sensory modality.

Various neuroimaging studies, using fMRI or positron emission tomography, have regularly reported strong amygdala involvement in the processing of emotion. The amygdala is involved whether the stimuli are visual, for example in processing fearful and other facial expressions (Breiter et al., 1996), or auditory, for example in processing fearful, happy, and sad vocalizations (Morris et al., 1999). Another study showed that the amygdala became activated when participants saw untrustworthy faces while the orbitofrontal cortex became more activated during explicit judgments of trustworthy faces (Winston et al., 2002). Given that the amygdala is involved with the processing of emotional stimuli, it could very well be *the* critical structure responsible for quickly and effectively evaluating information related to trust judgments. The orbitofrontal cortex, on the other hand, may be associated with analytical evaluation of trust information.

The orbitofrontal cortex comprises the ventromedial prefrontal cortex and the lateral-orbital prefrontal cortex. This area is reciprocally connected with the amygdala. A wide range of behaviors are associated with the orbitofrontal cortex, including social reasoning and social and emotional decision making. This brain area became a focus of both increased neurophysiological research and popular interest, given the well-known case of Phineas Gage, a railroad worker who suffered a localized brain lesion following injury by an iron rod.

Gage reportedly used rude, socially inappropriate language and was no longer able to function in society following his injury. Gage was unable to conform to the rules of etiquette because he was no longer concerned about the *face* of others. Virtual reconstruction of his brain showed that the rod had damaged the orbitofrontal cortex (Damasio, 1994). Damage to the orbitofrontal cortex leads to deficits in planning behaviors, and radical and inappropriate changes in social behavior, including outbursts of anger, while intellectual abilities are spared (Damasio, 1994). It may be that this area, at least in part, is also concerned with maintaining and monitoring social rules.

A component of the orbitofrontal cortex, the medial prefrontal cortex, has been reported to be associated with task processes that require theory of mind (Adolphs, 2003), particularly those that engage emotions and feelings that accompany social interactions. This area is active, for instance, during the appreciation of humor (Goel and Dolan, 2001), embarrassment (Berthoz et al., 2002), viewing of erotic stimuli (Karama et al., 2002), and elicitation of moral emotions (Moll et al., 2002).

In general, damage to orbitofrontal cortex results in an impaired ability to make decisions that require feedback from social or emotional cues and instead is characterized by an overreliance on perceptual information (Damasio, 1994). Reports of orbitofrontal lesions led to the development of the somatic marker hypothesis, in which emotion and feeling play an important role in decision making. It was proposed that this region of the brain provides access to "feeling states" associated with past experiences and that these states are recalled during similar experiences in which a decision will be made

(Damasio, 1994; Bechara et al., 2005). The somatic marker hypothesis proposes that the amygdala and ventromedial areas of the brain are considered to be emotional inducers. Primary inducers are learned or innate emotional stimuli that are encoded with a positive or negative emotional "tag" by the amygdala. Secondary inducers are stored emotional stimuli (i.e., the primary inducer) that can be recalled by the ventromedial cortex. Both primary and secondary inducers cause changes in bodily state that are subsequently translated into subjective feelings. These feelings then guide our decision making.

11.4.3.3 Using Arousal as a Measure of Emotion Perhaps the most measurable changes that occur within the human body, as a result of the use of appropriate etiquette, may be arousal. Arousal is a general physiological stimulation within the body at a particular point in time caused by inside or outside causes. The idea that arousal affects cognition was detailed in the two-factor theory of emotion. The first factor of this theory proposes that the body produces *nonspecific* physiological changes that can be taken as a measure of the intensity of arousal. The second factor of this theory proposes that the experience of feeling is an individual's attempt to attribute the arousal to a specific cause. This theory is consistent with recent neurological findings which suggest that arousal responses to environmental cues are first processed in brain areas such as the amygdala and the orbitofrontal cortex, and then create a visceral response in the body, while the anterior cingulate cortex evaluates its meaning.

To show that emotion plays a key role in decision making, Bechara et al. (1987) developed what is now known as the Iowa gambling task. Participants have to choose 100 cards from four decks. Decks A and B are high-risk decks because payoffs and costs are both high. In contrast, decks C and D have a low payoff, but also low costs. In the long run, decks C and D emerge as the advantageous decks. Behaviorally, healthy participants choose the advantageous decks consistently after about 20 cards. Patients with damage to the ventromedial area prefer the disadvantageous decks throughout a test session. Galvanic skin conductance can be used as a measure of emotional arousal. Bechara et al. (2005) found that healthy participants performing this task displayed a distinct galvanic skin response when choosing cards from high risk decks, but patients with ventromedial damage showed no such response. These results have been replicated using patients with amygdala lesions.

Bechara et al. (1997) provide a detailed explanation for this effect. When asked what they know about the game at different stages, participants show increasing awareness of the optimal playing rules as skin conductance response increases. The researchers divided the experiment into four phases including a prepenalty (no losses incurred), prehunch (no preferences for decks), hunch (preferences for certain decks), and conceptual phase (knowledge of good and bad decks). Healthy participants had no skin response in the prepenalty phase, but showed an increasing skin conductance response in the prehunch and hunch phases of the experiment. No skin conductance response was found for impaired patients. These findings have important implications for etiquette research. An FTA directed at a person can produce a physiological and emotional response. The severity of the physiological and emotional response will depend on the strength of the FTA. Severe FTAs have the potential to produce large emotional responses, and minor FTAs have the potential to produce small emotional responses. Politeness strategies can mitigate this effect, but any residual FTA that has not been addressed can be measured within the person. This has been referred to as residual arousal. The complete arousal state after a certain FTA has been presented could be an indicator of the severity of that threat. Further implications are that measurement of both emotion and behavioral decisions can serve as an indicator for the appropriateness of a politeness strategy.

11.4.4 The Application of Neuroergonomics to Etiquette

There are relatively few studies that directly examine the effects of dimensions of human–computer etiquette (such as politeness and interruptiveness) on neural measures reflecting arousal and emotion (see Chapter 4 by Wu et al. and Chapter 12 by Dorneich et al.) Measuring the person's cognitive state and determining what it means is still relatively new territory. As our review of neuroimaging studies illustrates, the relationship between emotions and neural responses observed in the brain is a complex one, without simple one-to-one correspondences. For example, showing an emotional stimulus to a participant does not lead to a simple increase in activity in the network of areas subserving the processing of emotional stimuli. Rather, the brain activation patterns observed depend upon the nature of the stimulus, the task that participants has to perform, and perhaps their strategy. Even distinguishing one emotion versus another is not a simple feat.

Before neuroergonomic measures can be used to provide effective guidance to etiquette-sensitive automated systems, we need to better understand the relationship between these variables in human communication paradigms.

11.5 Neuroergonomic Applications with Human Etiquette

Neuroergonomic guided, etiquette sensitive applications may be a powerful way to enhance many systems and devices that assist us in our daily lives. Systems that exhibit appropriate etiquette may increase the potential for healthy relationships to develop between technology and humans and, moreover, benefit individual and team performance.

In-vehicle navigation devices may be one candidate for such enhancements. While such devices are designed to lead a driver from point A to point B, this goal is not always accomplished without frustration and anger. Measuring the driver's emotion, and evoking an appropriate etiquette response may result in an improved driving experience. Research in this direction has already explored the effects of matching the voice of a navigation device to the emotional state of the driver (Nass et al., 2005). When they were matched, drivers experienced fewer accidents, attended more to the road, and spoke more to the device. If the emotional state of the user could be measured in real time, navigation devices could base their etiquette behavior on the state of the individual driver.

A second device that might be improved by the incorporation of neuroergonomics and etiquette strategies is adaptive cruise control. A critical issue for this technology is the transfer of control back and forth between the driver and the automation. Appropriate etiquette may provide a solution by warning the driver when the transfer occurs and doing so politely, or impolitely (urgently) if the situation warrants it.

A third application may be in automated training coaches (see Chapter 5 by Johnson and Wang and Chapter 9 by Bickmore). Given the measurement of a user's performance, these coaches could give feedback, depending on the level of engagement by the user. Users who need encouragement could receive a response from the automated system (e.g., "You're halfway there. Don't give up now!"). Users who are too engaged and need to calm down may also receive an appropriate warning (e.g., "Please slow down"). Such encouraging coaches may eventually be adapted to individual needs by introducing various coaching styles tailored to different people. Adaptive automation based on neuroergonomic measures can ensure that politeness and etiquette are used at appropriate moments (e.g., "Look out!" versus "Would you like a cup of tea?").

A fourth important application of neuroergonomics and etiquette may be robotics (see Chapter 10 by Zhang, Zhu, and Kaber). Advanced robots have already been developed for use in the home and to assist the elderly. Service robots need to be designed to foster trust from the users. For example, a rescue robot may need to perform small talk with a wounded person while rescue workers clear out the rubble and free the person. The small talk can be based on measured arousal and emotion from measures of face recognition, heart rate, and galvanic skin response (Breazeal and Picard, 2006). Other examples include robots that live in the homes of the elderly. If people are to rely on the robot to complete tasks, the robot needs to be trustworthy, and they need to trust the robot. Neuroergonomic measures could be useful in determining how far along this trust development has progressed, and to select appropriate etiquette strategies for deepening that relationship.

Encouragement, sympathy, and affective support in general would benefit a number of domains such as assistive technologies, medical communities, rescue robotics, and education. These developments are already under way with the design of relational robots that help children learn while measuring and responding to their emotional needs.

11.6 Conclusions

This chapter provides a review of recent neurophysiological finding, and explores the possibility of measuring emotional responses through neurophysiological indicators so that we may create systems that can adjust their behavior accordingly. Finally, we have suggested several applications in which neurophysiological indicators can be used to inform the responses of several different types of automated devices that we use in our daily lives, enabling them to behave more like polite and cooperative collaborators than rude technological tyrants.

References

Adolphs, R. (2003, March). Cognitive neuroscience of human social behavior. *Nature Reviews Neuroscience, 4*, 165–178.

Asimo Honda (2010). Honda's ASIMO robot to conduct the Detroit Symphony Orchestra. In Asimo: The world's most advanced humanoid robot. Retrieved September 10, 2010, from http://asimo.honda.com/.

Axelrod, R. and Hamilton, W. D. (1981). The evolution of cooperation. *Science, 211*, 1390–1396.

Baron-Cohen, S. (2000). Theory of mind and autism. International Review of Research in Mental Retardation, 23, 169–184.

Bechara, A., Damasio, H., Tranel, D., and Damasio, A.R. (1997). Deciding advantageously before knowing the advantageous strategy, *Science, 275*, 1293–1295.

Bechara, A. and Damasio, A. R. (2005). The somatic maker hypothesis: A neural theory of economic decision. *Games and Economic Behavior, 52*(2), 336–372.

Berg, J., Dickhaut, J., and McCabe, K. (1995). Trust, reciprocity, and social history. *Games and Economic Behavior, 10*, 122–142.

Berthoz, S., Armony, J. L., Blair, R. J. R., and Dolan, R. J. (2002). An fMRI study of intentional and unintentional (embarrassing) violations of social norms. *Brain, 125*, 1696–1708.

Breazeal, C. and Picard, R. (2006). The role of emotional-inspired abilities in relational robots. In R. Parasuraman and M. Rizzo (Eds.), *Neuroergonomics: The Brain at Work*: Oxford University Press, New York.

Breiter, H. C., Etcoff, N. L., Whalen, P. J., Kennedy, W. A., Rauch, S. L., Buckner, R. L., Strauss, M. M., Hyman, S. E., and Rosen, B. R. (1996). Response and habituation of the human amygdale during visual processing of facial expression, *Neuron, 17*, 875–887.

Brown, P. and Levinson, S. (1987). *Politeness: Some Universals in Language Usage*. Cambridge, U.K.: Cambridge University Press.

Camerer, C. F. and Weigelt, K. (1988). Experimental tests of a sequential equilibrium reputation model. *Econometrica, 56*, 1–36.

Carter, C. S. (1998). Neuroendocrine perspectives on social attachment and love. *Psychoneuroendocrinology, 23*, 779–818.

Coleman, J. S. (1990). *Foundations of Social Theory*. Cambridge, MA: Harvard University Press, pp. 177–179.

Damasio, A. R. (1994). *Descartes' Error Emotion, Reason, and the Human Brain.* New York: Putnam.

de Visser, E. J., Shaw, T., Rovira, R., and Parasuraman, P. (2009). Could you be a little nicer? Pushing the right buttons with automation etiquette. In *Proceedings of the International Ergonomics Association Conference*, Beijing, China.

Delgado, M. R., Frank, R. H., and Phelps, E. A. (2005). Perceptions of moral character modulate the neural systems of reward during the trust game. *Nature Neuroscience, 8 (11)*, 1611–1618.

Dolan, R. J. (2002). Emotion, cognition, and behavior. *Science, 298*, 1191–1194.

Fehr, E., Kirchsteiger, G., and Riedl, A. (1993). Does fairness prevent market clearing? An experimental investigation. *The Quarterly Journal of Economics, 108 (2)*, 437–459.

Gazzaniga, M. S., Ivry, R. B., and Mangun, G.R. (2002). *Cognitive Neuroscience: The Biology of Mind (2nd ed.).* New York: W.W. Norton & Company, Inc.

Goel, V. and Dolan, R. J. (2001). The functional anatomy of humor: Segregating cognitive and affective components. *Nature Neuroscience, 4*, 237–238.

Halgren, E., Raij, T., Marinkovic, K., Jousmaki, V., and Hari, R. (2000). Cognitive response profile of the human fusiform face area as determined by MEG. *Cerebral Cortex, 10*, 69–81.

Hancock, P. A. and Szalma, J. L. (2008). *Performance under Stress*: Ashgate Publishing, Ltd.

Hancock, P. A. and Warm, J. S. (1989). A dynamic model of stress and sustained attention. *Human Factors, 31* (5), 519–537.

Haxby, J. V., Hoffman, E. A., and Gobbini, M. I. (2000). The distributed human neural system for face perception. *Trends in Cognitive Sciences, 4*, 223–233.

Insel, T. R. and Shapiro, L. E. (1992). Oxytocin receptor distribution reflects social organization in monogamous and polygamous voles. *Proceedings of the National Academy of Sciences USA, 89*, 5981–5985.

Karama, S., Lecours, A. R., Leroux, J. M., Bourgouin, P., Beaudoin, G., Joubert, S., and Beauregard, M. (2002). Areas of brain activation in males and females during viewing of erotic film excerpts. *Human Brain Mapping, 16*, 1–13.

King-Casas, B., Tomlin, D., Anen, C., Camerer, C. F., Quartz, S. R., and Montague, P. R. (2005, April). Getting to know you: Reputation and trust in a two-person economic exchange. *Science, 308*, 78–83.

Kosfeld, M., Heinrichs, M., Zak, P. J., Fischbacher, U., and Fehr, E. (2005). Oxytocin increases trust in humans. *Nature, 435*, 673–676.

Lee, J. and See, K. (2004). Trust in automation: Designing for appropriate reliance. *Human Factors, 46*(1), 50–80.

Lerner, J. S. and Keltner, D. (2000). Beyond valence: Toward a model of emotion-specific influences on judgment and choice. *Cognition and Emotion* 14, no. 4 (2000): 473–93.

McCarthy, G. (1999). Physiological studies of face processing in humans. In M. S. Gazzaniga (Ed.), *The New Cognitive Neurosciences* (pp. 393–410). Cambridge, MA: MIT Press.

McClure, S. M., Berns, G. S., and Montague, P. R. (2003). Temporal prediction errors in a passive learning task activate human striatum. *Neuron, 38*, 339–346.

Miller, C. A. (2004). Human-computer etiquette: Managing expectations with intentional agents. *Communications of the ACM, 47*(4), 30–34.

Miller, C. A., Wu, P., and Funk, H. (2007, August 27–28). A computational approach to etiquette and politeness: Validation experiments. Paper presented at the *Proceedings of the First International Conference on Computational Cultural Dynamics*, College Park, MD.

Moll, J., de Oliveira-Souza, R., Eslinger, P. J., Bramati, I. E., Mourão-Miranda, J., Andreiuolo, P. A., and Pessoa, L. (2002). The neural correlates of moral sensitivity: A functional magnetic resonance imaging investigation of basic and moral emotions. *Journal of Neuroscience, 22*, 2730–2736.

Morris, J. S., Scott, S. K., and Dolan, R. J. (1999). Saying it with feeling: Neural responses to emotional vocalizations. *Neuropsychologia, 37*, 1155–1163.

Nass, C. (2004). Etiquette equality: Exhibitions and expectations of computer politeness. *Communications of the ACM, 47*(4), 35–37.

Nass, C., Steuer, J., and Tauber, E. R. (1994). Computers are social actors. Paper presented at the *Proceedings of the SIGCHI Conference on Human factors in Computing Systems: Celebrating Interdependence*. Boston: MA, 72–78.

Nass, C., Jonsson, I., Harris, H., Reaves, B., Endo, J., Brave, S., and Takayama, L. (2005). Improving automotive safety by pairing driver emotion and car voice emotion. In *CHI '05 Extended Abstracts on Human Factors in Computing Systems*. Portland, OR: ACM (pp. 1973–1976).

O'Doherty, J. P., Dayan, P., Schultz, J., Deichmann, R., Friston, K., and Dolan, R. J. (2004). Dissociable roles of ventral and dorsal striatum in instrumental conditioning. *Science, 304 (5669)*, 452–454.

Packard, M. G. and Knowlton, B. J. (2002). Learning and memory functions of the basal ganglia. *Annual Review of Neuroscience, 25*, 563–593.

Pagnoni, G., Zink, C. F., Montague, P. R., and Berns, G. S. (2002). Activity in human ventral striatum locked to errors of reward prediction. *Nature Neuroscience, 5*, 97–98.

Parasuraman, R. (2003). Neuroergonomics: Research and practice. *Theoretical Issues in Ergonomics Science, 4,* 5–20.

Parasuraman, R. and Miller, C. A. (2004). Trust and etiquette in high-criticality automated systems. *Communications of the ACM, 47*(4), 51–55.

Parasuraman, R. and Riley, V. (1997). Humans and automation: Use, misuse, disuse, abuse. *Human Factors, 39*(2), 230–253.

Parasuraman, R. and Rizzo, M. (2007). Introduction to neuroergonomics. In R. Parasuraman and M. Rizzo (Eds.), *Neuroergonomics: The Brain at Work* (pp. 3–12). New York: Oxford University Press.

Parasuraman, R. and Wilson, G. F. (2008). Putting the brain to work: Neuroergonomics past, present, and future. *Human Factors*, in press.

Pizzagalli, D., Regard, M., and Lehmann, D. (1999). Rapid emotional face processing in the human right and left brain hemispheres: An ERP study. *Neuroreport, 10*, 2691–2698.

Poldrack, R. A. (2001). Interactive memory systems in the human brain. *Nature, 414*, 546–550.

Seymour, B., O'Doherty, J. P., Dayan, P., Koltzenburg, M., Jones, A. K., Dolan, R. J., Friston, K. J., and Frackowiak, R. S. (2004). Temporal difference models describe higher-order learning in humans. *Nature, 429 (6992)*, 664–667.

Shohamy, D. (2004). Cortico-striatal contributions to feedback-based learning: converging data from neuroimaging and neuropsychology. *Brain, 127*, 851–859.

Takahashi, T., Ikeda, K., Ishikawa, M., Tsukasaki, T., Nakama, D., Tanida, S., and Kameda, T. (2004). Social stress-induced cortisol elevation acutely impairs social memory in humans. *Neuroscience Letters, 363*, 125–130.

Takahashi, T. (2005). Social memory, social stress, and economic behaviors. *Brain Research Bulletin, 67*, 398–402.

Tricomi, E. M., Delgado, M. R., and Fiez, J. A. (2004). Modulation of caudate activity by action contingency. *Neuron, 41*, 281–292.

Winston, J. S., Strange, B. A., O'Doherty, J., and Dolan, R. J. (2002). Automatic and intentional brain responses during evaluation of trustworthiness of faces. *Nature Neuroscience 5*, no. 3: 277–83.

12

Etiquette Considerations for Adaptive Systems that Interrupt

Cost and Benefits

MICHAEL C. DORNEICH, SANTOSH MATHAN, STEPHEN WHITLOW, PATRICIA MAY VERVERS, AND CAROLINE C. HAYES

Contents

74th Rule When another speaks, be attentive yourself and disturb not the audience. If any hesitate in his words, help him not, nor prompt him without it being desired; interrupt him not, nor answer him until his speech be ended.

—George Washington's Rules of Civility*

12.1 Introduction

It is not generally considered polite to interrupt. However, etiquette is not simply about politeness but also appropriateness. Thus, while etiquette dictates that it is inappropriate, counterproductive, or even dangerous to interrupt someone with irrelevant information in the middle of a critical task requiring great concentration, it also allows that it is appropriate and necessary to interrupt with crucial information. For example, it is inappropriate to interrupt an airline pilot during landing with the baseball score, but entirely appropriate to interrupt with the information that the landing gear has malfunctioned.

A good human assistant can be invaluable in helping a decision maker to manage interruptions. Consider, for example, the value of a good administrative assistant (or field assistant) in deferring noncritical requests, and passing through urgent and pertinent requests, sometimes drawing attention to them through interruptions. Similarly, computer assistants hold the potential to improve decision makers' performance, either by (1) minimizing interruptions with noncritical information or requests, or (2) redirecting attention (e.g., interrupting) at critical junctures to necessary information.

Unfortunately, computer assistants typically lack the savvy of human assistants to know when it is appropriate to interrupt. The result is often computer assistants that appear to users like small children who interrupt at inappropriate times. This lack of savvy stems both from computers' inability to understand when the user is "busy," and what constitutes an appropriate interruption. Traditionally, computer assistants attempt to infer the user's current workload, task, and needs based on the user's current actions, and the computer's understanding of the task and environment. Such approaches are often prone to

* Washington, George (2003). George Washington's Rules of Civility, fourth revised Collector's Classic edition, Goose Creek Productions, Virginia Beach, Virginia.

error when they incorrectly guess the user's current cognitive state and workload, and end up annoying more than assisting them.

Recently, however, researchers have been exploring more direct means for assessing humans' cognitive state using physical and neurophysiological sensors to assess brain activity, heart rate, skin conductance (sweating), and other indicators. This raw sensor information is then processed to derive an assessment of the human's cognitive state. The first challenge is to determine whether such sensor-informed methods can correctly classify the human's internal state; the second is to determine whether this information can be used to enable a computer system to adapt its behavior in a useful way. In earlier work, we addressed the first challenge by developing and testing a cognitive state classifier that could assess when mobile users were "busy" by monitoring a combination of indicators of brain activity and heart rate (Dorneich et al., 2006; 2007). "Busy," in this case, means users' cognitive workload was so high that performance on one or more tasks may start to suffer. In this work, it is our goal to address the second challenge: to determine whether we could incorporate the cognitive state classifier in two adaptive systems aimed at enhancing task performance for mobile dismounted soldiers. An *adaptive system* is one in which humans and computers work together on joint tasks, and either can each initiate changes as to which tasks are managed by the human and which by the computer (Scerbo, 1996). These systems will literally read the "hearts and minds" of soldiers.

The specific way in which these adaptive systems assist soldiers is in managing interruptions.

For soldiers and commanders operating in the field, interruptions are a way of life. In high-stress times they must juggle multiple high-criticality tasks simultaneously, such as navigating through an urban war zone while avoiding enemy fire, along with less immediate but equally necessary tasks, such as coordination with other friendly units and keeping tabs on the evolving mission plan. Survival and success require not only that soldiers complete mission tasks, but also manage interruptions, and filter information in an effective way.

To help them manage these interruptions and maximize performance during very "busy" high workload periods, we designed and built two adaptive systems. The first, the Communication Scheduler, assists soldiers by deciding *when not to interrupt*, while the second, the Tactile Navigation Cueing System, assists by deciding *when to interrupt* with critical information. Both systems are designed to help dismounted soldiers manage critical information as they carry out multiple, simultaneous tasks that are carried out in an outdoor field, in situations mimicking urban warfare settings. Both systems use EEG (brain) and ECG (heart) sensors as the basis of a cognitive state classifier to identify when assistance is most needed.

The decisions of when to interrupt and when *not* to interrupt are critical etiquette issues that are equally relevant in interactions between humans as they are in interactions between humans and computers. In other chapters in this volume (Wu et al.), researchers have examined etiquette in terms of Brown and Levinson's model (1987), which focuses on etiquette in spoken and text communications as a means to mitigate the *face threat*, or the imposition placed on the listener by a speaker through the act of communicating. Brown and Levinson additionally state that "politeness … makes possible communication between potentially aggressive parties" (1987).

In this work, our focus is not on face threats or verbal communications* per se, but on the impact that interruptions have on task performance, including the complex trade-offs between the benefits and costs that result under various conditions. We posit that etiquette guidelines associated with interruptions in human social interactions are based on a cooperative desire to maximize the performance of a group who share a common set of tasks and goals. We found that maximizing performance is complex and highly situation dependant. It requires constant monitoring of the human's cognitive state and the ability to adapt as circumstances change. This nuanced understanding, which may allow either computers or humans to judge when to interrupt or hold back, is an important part of etiquette. In this view, etiquette is not just a way of mitigating potential hostilities in a competitive group, it is also a way of maximizing effectiveness in a cooperative group, regardless of whether it is a group of humans, or a group of humans and computers working together.

* Some of the interruptions in our work are nonverbal, such as the buzzing of a vibrating pad worn against the body.

12.2 Related Work

Overall, the two systems described in this chapter are designed to manage interruptions in a highly dynamic, multitasking environment. This section provides background on interruption management and neurophysiological adaptive systems, and finally introduces some areas where etiquette considerations impact the design of such systems.

12.2.1 The Need to Manage Interruptions

As more technology is developed to assist people, it has become increasingly difficult to manage the constant flow of interruptions from computer tasks that are intended to help us. Computer and communications systems are more often than not designed with little consideration of social conventions or the social impact of their actions (Fogarty, Hudson, and Lai, 2004). These constant interruptions are not only irritating, but the very tools introduced to increase productivity can actually rob us of it. For instance, common office workers spend up to 25% of their workday recovering from interruptions or switching between tasks (Spira, 2005), and office workers are interrupted or distracted on average once every 3 minutes (Zeldes, Sward, and Louchheim 2007). Common interruptions include telephones rings, e-mail chimes, pop-up calendar reminders, and instant messaging alerts; all demand attention irrespective of what the person is doing currently, and often interrupt at inopportune times (Chen, Hart, and Vertegaal, 2007). Inappropriate interruptions can increase errors, frustration, and stress, and reduce efficiency and decision quality (Iqbal, Adamcyzk, Zheng, and Bailer, 2005; McFarlane and Latorella, 2002; Gillie and Broadbent, 1989; Chen and Vertegaal, 2004). There is a general desire to create automated systems that are less frustrating to users by endowing them with the same courtesies exhibited by human colleagues. McFarlane and Latorella (2002) concluded that interruption management is an important human–computer interaction issue that will only grow in importance and ubiquity.

The cost of inappropriate interruptions can be particularly high in high workload, safety-critical domains such as the military, disaster management, and medical emergencies, where errors can have life or death consequences (Mathan, Whitlow, Dorneich, Ververs, and Davis,

2007). For example, McFarlane and Latorella (2002) describe the impact of task scheduling and interruptions on overall system performance of the AEGIS Weapons System, which coordinates the activities of over 35 sailors simultaneously and generates many different alerts and interruptions. An exercise using the system in the mid-1990s revealed that the system was sending alerts to some human operators every 11.5 seconds, each alert requiring a minimum of 5 seconds to address.

Many systems have been built to manage interruptions. One key element is some consideration of the user's current tasks or workload to determine how "busy" or interruptible a person is at a given time. The ability to predict someone's interruptibility would allow the designers of automated systems to take advantage of periods of high interruptability to display incoming alerts, suggest the user attend to different information, or switch tasks, and to protect periods of low interruptability by minimizing disruptions (Feigh and Dorneich, in preparation). However, system designers have historically found it difficult to accurately model user interruptibility. Often user workload could only be measured via indirect means, such as monitoring interactions with the system (e.g., keystrokes, button pushes), monitoring communications bandwidth, assessing performance, or modeling task load. Without accurate models of user tasks and workload, information management systems were limited in how effectively they could manage interruptions. Such systems often failed to gain user acceptance because they either filtered to much or too little information (Parasuraman and Miller, 2004). Improved interruption management continues to be an area of interest and concern in the field of human–computer interaction (Hudson et al., 2003; Adamczyk et al., 2005).

12.2.2 The Need to Understand User Context and Workload

There has been research on ways to make systems more aware of the user's context in order to better time and tailor interruptions (Chen and Vertegaal, 2004; Mathan et al., 2007; Fogarty, Hudson, and Lai, 2004; Wickens and Hollands, 2000). In order for an adaptive system to decide when to interrupt, it must have some sense of what is happening. In what tasks is the user engaged? How important are those tasks? How busy or overloaded is the user (e.g., what is his or her cognitive workload?)? This knowledge can be derived in many ways.

Computer systems often assess users' cognitive workload through indirect measures such as their activities (Horvitz, 1999), focus of attention (e.g., Duchowski, 2003), performance (e.g., Hancock and Chignell, 1987; Scerbo, 1996), and interactions with devices (e.g., Shell, Vertegaal, and Skaburskis, 2003). Inferences based on observations of these factors are often used to predict whether a user is engaged and if it is an appropriate time to interrupt.

There are a number of limitations of basing cognitive state estimation on indirect, model-based methods. First, these approaches often rely on the instrumentation of the task environment and are well suited to sedentary work environments (e.g., offices, process control plants). Second, models work best when the contributors of user workload are predictable and well understood. However, task demands can change in unpredictable ways in many complex task environments (e.g., military). Third, model creation in complex task domains is very time consuming and expensive, and these recurring costs grow with the complexity of environment and variability of individuals. Therefore, in task domains where sources that contribute to workload are unpredictable and difficult to track, where users vary widely in experience, and the task environment is complex, a more direct measure of the user's cognitive workload may be needed. This is especially important in high workload domains where the cost of inappropriate interruptions can be extreme (Mathan et al., 2007).

More recently, some lines of research have begun to focus on methods to directly measure cognitive workload and attention via physiological and neurophysiological sensors (Pope, Bogart, and Bartolome, 1995; Prinzel, Freeman, Scerbo, Mikulka, and Pope, 2000; Dorneich et al., 2007). Neurophysiological- and physiological-based assessment of cognitive state has been captured in several different ways, including measures of heart activity via electrocardiogram (ECG), and brain activity via electroencephalogram (EEG) and functional near-infrared (fNIR) imaging.

One of the most promising approaches is the use of EEG to measure cognitive state. EEG measures the electrical activity on the scalp associated with neuronal firing in the brain. It is a fast signal, with excellent temporal resolution, and therefore is a moment-to-moment measure. The signal is typically analyzed by decomposing the signal in the frequency domain and looking for relationships to constructs associated with information processing. For instance, working memory has been associated with frontal theta waves, where attention has been associated with midline alpha waves. As the "gold standard" for providing high-resolution temporal indices of cognitive activity, EEG has been used in the context of adaptive systems. Research has shown that EEG activity can be used to assess a variety of cognitive states that affect complex task performance, such as working memory (Gevins and Smith, 2000), alertness (Makeig and Jung, 1995), engagement (Pope, Bogart, and Bartolome, 1995), executive control (Garavan, Ross, Li, and Stein, 2000), and visual information processing (Thorpe, Fize, and Marlot, 1996). These findings point to the potential for using EEG measurements as the basis for driving adaptive systems that demonstrate a high degree of sensitivity and adaptability to human operators in complex task environments. For instance, Prinzel and colleagues (2000) used measures of workload to drive an adaptive system to reallocate tasks during periods of low human cognitive engagement, as derived from measurements of EEG. In earlier work, we demonstrated that adaptive scheduling of communications based on cognitive state assessment of the readiness to process information resulted in a twofold increase in message comprehension and situation awareness (Dorneich et al., 2006). These results highlight the potential benefits of a neurophysiologically triggered adaptive automation. Many of the limitations of model-based approaches are alleviated with direct estimates of the operator's cognitive state. Complex task models based on indirect measures are not required to estimate cognitive workload. In fact, with reliable, real-time measure of the cognitive state, adaptive systems can be much more proactive and aggressive in managing user interruptions. However, such systems must be designed with consideration of human–computer interaction etiquette, lest they fail to gain user acceptance because the design fails to recognize the costs associated with any automated intervention, despite the benefits.

12.3 Two Adaptive Systems

12.3.1 Introduction

We performed evaluations on two adaptive systems designed to assist the dismounted soldier: the Communications Scheduler and the

Tactile Navigation Cueing System. The Communications Scheduler minimized interruptions during high workload times, while the Tactile Navigation Cueing System purposely interrupted as the soldier navigated a mine field to direct him or her away from dangerous areas. We examined the impact of these systems on soldiers' performance during high and low workload times, and we examined both the benefits and the costs. Our results both quantify intuitions about interruptions in human interactions, and provide insights for the design and use of intelligent computer assistants.

Both systems are portable and mounted on the body so that dismounted soldiers can carry the system with them in the field. The computer was carried in a backpack, and sensors and other equipment were mounted on the head and body. Both systems monitor the soldier's cognitive state using EEG and ECG sensors. The EEG sensors are incorporated into a special cap that participants wore. The ECG sensors were glued to the torso with a medical adhesive patch. All sensors were connected wirelessly to the computer in the backpack. These sensors are used to assess the soldier's cognitive workload (high or low). We have tested this approach for monitoring cognitive state for users in simulated environments, and mobile users in field environments. We were able to achieve overall classification accuracy into the 90% range (Dorneich et al., 2006).

We found that there is typically a cost associated with *interruptions*, but that this cost may be repaid many times over under some circumstances—for example, if that interruption is a timely warning that you are about to step on a mine. Conversely, there is also a cost associated with *minimizing interruptions*; minimizing interruptions may be necessary at critical times in order to achieve focus. However, the cost of doing so may be a loss of situation awareness about other tasks, reduced performance on those tasks, or other negative consequences. The costs and benefits of computer decision support tools that either create or minimize interruptions must be carefully weighed when designing such tools, and deciding when to and where to apply them.

12.3.2 Communications Scheduler

In a military field operations, commanders are expected to carry out multiple tasks simultaneously, and to perform them all well. Their lives, and the lives of those whom they command, depend on it. In an urban warfare situation, the members of a platoon are often spread out across a location, working towards a joint objective. They may not always be able to see each other, but they will be in radio contact. The commander must keep track of the positions of his or her own troops, other friendly troops, plan the next move, and watch for and count civilians and enemy soldiers and report those counts to others. Additionally, the commander may be interrupted sporadically by anything from enemy fire to requests from headquarters for information. Communication about most of these activities will occur through the radio. Some of the radio messages may be directed to a particular commander, while others may concern other units. However, it is still important for the commander to monitor such messages in order maintain awareness of where those other friendly units are, in case one needs to call on them for help and to avoid firing on them.

The Communications Scheduler, shown in Figure 12.1, is designed to help soldiers manage their workload and maintain performance on high priority tasks. It does so by detecting when their workload becomes very high (using the EEG cap and ECG patch), and diverting low-priority radio messages to their PDA as text messages (see Figure 12.2). High priority messages continue to be delivered to them as radio messages. The text messages sent to the PDA are available immediately to the soldier if they want to look

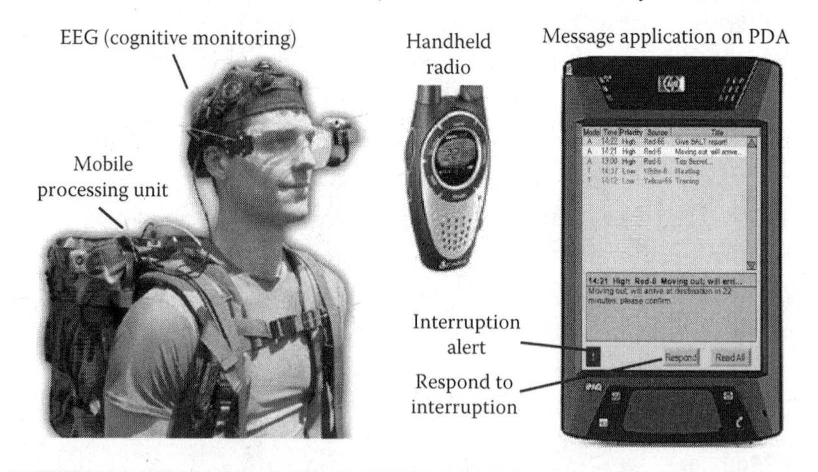

Figure 12.1 The Communications Scheduler includes an EEG-based cognitive monitoring system to assess workload, a hand-held radio for aural messages, and the hand-held PDA for text messages.

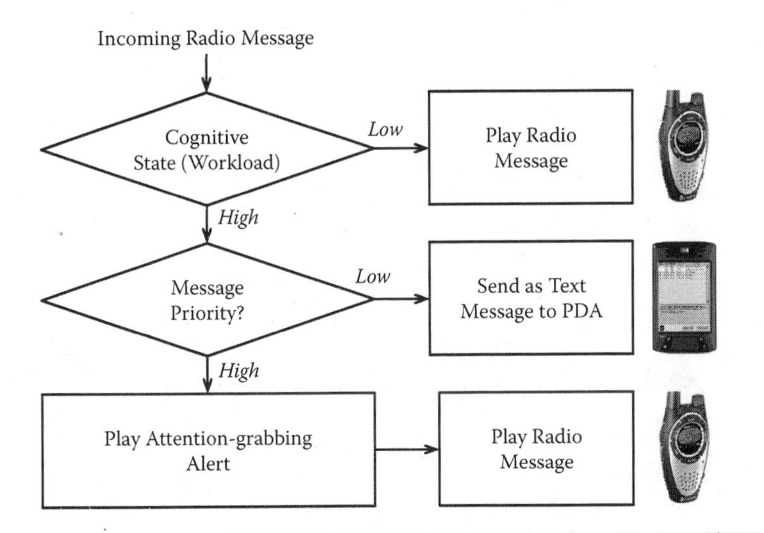

Figure 12.2 If user's workload is high, then the Communications Scheduler interrupts with an alert if message priority is high or defers the message as text to the PDA if the priority is low.

at them, but they do not have to. Most important is the fact that text messages are less salient or "attention grabbing" than radio messages. By converting the low priority messages to a less salient form, it allows soldiers to focus on the high priority messages during high workload periods.

To enable the Communication Scheduler to determine which messages to present as radio messages, and which to divert to the PDA, all messages had a priority assigned to them *a priori* by the scenario designer. The priority (or criticality) of messages is determined by many factors such as the source (e.g., your commander, another commander, a fellow soldier), the relevance of the content to the tasks at hand, etc. High priority items were typically mission-critical and time-critical. It was imperative that they should receive the commander's attention as soon as they arrived. Low priority messages were not critical, although they may still be important.

12.3.3 Tactile Navigation Cueing System

Sometimes the best way to improve a soldier's performance during a high workload time is to shield him or her from noncritical demands on his or her attention. In other situations, the best way to improve performance may be to manage one or more of the tasks for the soldier.

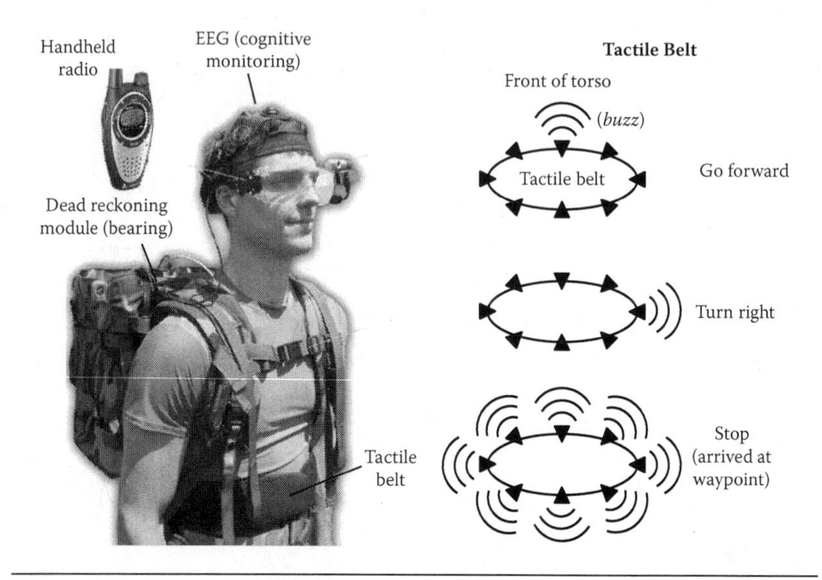

Figure 12.3 The Tactile Navigation Cueing System (TNCS). The figures on the left provide a top down view in which up is the front of the torso.

However, doing so may also require interacting with the soldier and potentially interrupting concentration on other important tasks. The Tactile Navigation Cueing System (TNCS) (see Figure 12.3) was designed to assist the soldier by taking over the task of navigating safely through a mine field during high workload periods. However to do so, it also needed to give directives (turn left, go straight, etc.) which may have distracted from other important tasks, such as avoiding enemy surveillance, and keeping track of the location and status of subordinate units. The essential question is, do the benefits of relieving the soldier of the navigation task outweigh the costs of interrupting to give directions?

In our experimental trials, the soldiers moved through a real field with fake mines and fake surveillance cameras; the mines and cameras were simply locations to be avoided in the field. The experimental scenario was designed to reflect a combat situation in which soldiers must quickly navigate a complex, unfamiliar, and dangerous route while simultaneously carrying out other tasks, such as keeping track of their own troops, watching for civilians, evading enemy soldiers, and avoiding enemy fire. Under such conditions, one could easily get distracted from avoiding mines, with tragic results.

The goal of the TNCS is to help soldiers safely navigate a mine field when workload is high. The equipment setup for the TNCS was very similar to that for the Communications Scheduler. Soldiers wore a computer in a backpack, a radio clipped to their shoulder, a PDA in their hand, an ECG pad glued to their trunk, and a hat with EEG sensors. The TNCS used the same cognitive state classifier that the Communication Scheduler used to detect when the soldier was experiencing low or high workload. Additionally, the computer was connected to a GPS system to determine their exact location, and soldiers wore a *tactor belt*: a belt with eight evenly spaced, vibrating pads. The soldiers wore the tactor belts around their waists so that the pads were in contact with their bodies. The GPS system supplied the Dead Reckoning Module with the information needed to guide the participant though a series of waypoints.

When the TNCS detected a high workload, the TNCS took over the task of navigating. The system defined a route between mines and around cameras as a series of straight lines between waypoints. As long as soldiers stuck closely to this route they would not trip mines or risk detection by surveillance cameras. The systems conveyed directions to the soldier through the tactor belt. The pads in the tactor belt vibrated to tell the soldier which direction to go. For example, if a pad on the left vibrates, that means "go left"; if the pad in front vibrates, "continue forward." The rate of firing the pads increased as the participant approached each waypoint. When a waypoint was reached, all the tactors fired simultaneously to indicate "stop." Then the system would begin to provide directions to the next waypoint, until the soldier reached the end of the route.

12.4 Evaluations

The Communications Scheduler and the Tactile Navigation Cueing System were evaluated in two field experiments to assess their appropriate use, potential benefits, and potential costs.

12.4.1 Experiment A: Communications Scheduler

This experiment investigated the costs and benefits of *minimizing* interruptions during field operations, under a variety of conditions. We did so by evaluating the Communication Scheduler's impact on multiple aspects of the mobile soldier's task performance during both high workload and low workload conditions. Our hypotheses were that:

1. During high workload times, use of the Communications Scheduler would enhance performance on high priority tasks when the soldier's work load is high, while not degrading performance on other tasks.
2. During low workload times, there may be costs associated with the Communications Scheduler that make it inappropriate to leave it on all the time.

12.4.1.1 Operational Scenario The evaluation was held outdoors in a field of roughly 6500 square meters that consisted of primarily open grassy areas with some tree cover and forest in other areas. Each participant played the role of a platoon leader (PL), and was part of a larger simulated military company. A company is comprised of roughly 62–190 soldiers. A platoon is comprised of roughly 16–44, grouped into squads of 9–10 soldiers each (Powers, 2010). The commander listened to messages from "soldiers" who were actually prerecorded audio messages, each with a different voice, rather than live confederates in the experiment. The computer played each message to the participant according to a script.

Each participant was the leader of the Red platoon, and his or her call sign was "Red-6." Each participant was responsible for managing three squads while reporting to his or her company commander (CO). The squad leaders' call signs were "Red-6-1," "Red-6-2," and "Red-6-3." Figure 12.4 illustrates the command hierarchy and gives

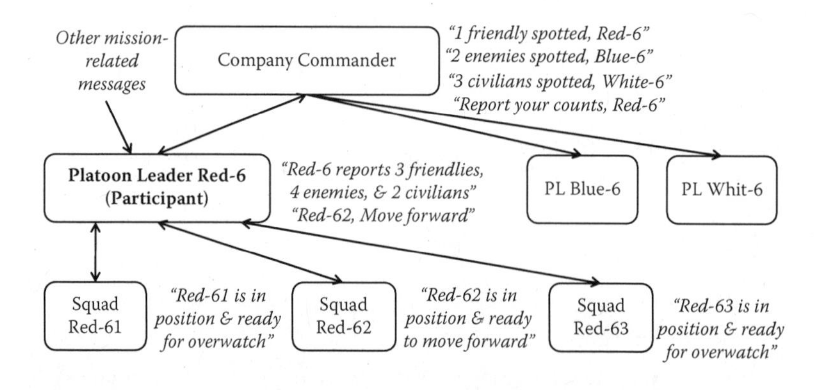

Figure 12.4 The command structure, and messages originating at each level.

examples of some of the task-related incoming messages heard by the participant, and outgoing messages from the participant.

12.4.1.2 Tasks Participants were given five tasks to perform simultaneously: navigation along a route, keeping count of the number of civilians and soldiers sighted, maneuver monitoring, and maintaining awareness of the overall situation. Additionally, there was an "interruption" task; while all radio messages involved in the tasks were interruptions, the interruption task was designed to be especially disruptive, requiring much concentration and focus. All the tasks were designed to resemble militarily relevant tasks to create experimental conditions as realistic as possible, while still permitting easy collection of performance metrics assessing the speed and accuracy in completing the tasks. Finally, these tasks allowed us to vary the participant's cognitive workload by varying the rate at which messages were delivered, or increasing the number and complexity of the tasks.

We asked soldiers to perform five tasks simultaneously because we had learned in prior studies that people can in fact manage many tasks simultaneously without becoming cognitively overloaded, especially if they are tasks at which they have become skilled. Additionally, we have observed that the more artificial the situation, the more difficult it becomes to "overload" a person interacting in that environment, possibly because artificial environments lack the detail, richness, and stimulation of a real environment. For example, we tested early versions of the Communications Scheduler in a simulated environment in which participants navigated in a "game" environment peopled by computer controlled characters. Workload was manipulated by varying the number of enemy soldiers shooting at the participant. We found that it was very hard to overload subjects in this environment; there had to be 30 or more characters shooting at subjects before it interfered with their ability to perform other tasks (Dorneich et al., 2004).

Additionally we have observed that the richer and more realistic the environment, the easier it becomes to overload subjects. For, example in later studies (subsequent to the work reported in this chapter) soldiers participating in a field experiment moved through a semiurban environment shooting soap "bullets" at each other, and it took relatively few simultaneous tasks to overload them enough to decrease task performance (Dorneich et al., 2007).

The context in the Communications Scheduler experiment, reported in this chapter, was relatively realistic, while not completely real; subjects moved through a real field, but the "people" with whom they interacted were only recordings. Our experience indicated that at least five simultaneous tasks would be required to seriously tax subjects' cognitive resources, and only when those tasks were made particularly difficult (for example, by increasing the rate at which the messages were delivered).

Following is a more detailed description of the five tasks used in the experiment:

1. *Navigation.* Participants were asked to walk along a simple, familiar, and well marked route during each trail. The participants were mobile during the task so as to approximate real field conditions; moving in the richness of a real environment requires far more attention and cognitive resources than sitting at a computer and moving through a simulated world. Additionally, we wanted to demonstrate that a system such as the Communications Scheduler could provide useful support to a mobile field soldier. We had already demonstrated that a combination of ECG (heart) and EEG (brain waves) could successfully measure the workload of a mobile soldier (Erdogmus et al., 2005); we were now putting the whole system together, and our aim was to assess whether useful support, based on physiological and neurophysiological workload measurements, could be provided to mobile soldiers in a field environment.
2. *Maintain radio counts.* A simulated Company Commander (CO) broadcast radio messages to the three platoon leaders about the number of civilians, enemy soldiers, and friendly soldiers sighted. Two of the platoon leaders (White-6 and Blue-6) were simulated only as radio voices. The participant in the experiment played the role of platoon leader "Red-6." The participant was asked to maintain a running total of civilians, enemies, and friendlies reported to "Red-6," while ignoring the counts reported to the other two platoon leaders. This task required participants to attend to all messages in sufficient detail to determine whether it was directed at them or other platoon leaders, and to keep counts in working memory until asked by

the commander to report the counts. (Once they had reported a count they could start again at zero). Performance on this task was measured by the number of correct counts reported by the participant for civilians, enemy, and friendly soldiers.

3. *Maneuver monitoring.* A common duty for a platoon leader is to orchestrate a series of maneuvers among the squads under his or her command. For example, in a *bounded overwatch*, the platoon moves towards its objective through a series of steps in which one squad moves, while the other two squads protect the moving squad. In our study, participants were asked to keep track of the status of all three (simulated) squads under their command. Each squad would radio a message to Red-6 once they were in position for the maneuver. Two squads would report "ready for overwatch," and one squad, "ready to move." Their reports would arrive in random order. Once all three were in position, the participant radioed an order commanding the "ready-to-move" squad to move forward. This task required the participant to keep track of each of the three squads' status in working memory until the all squads were in position. Performance on this task was measured by the participant's accuracy in sending the correct team forward (e.g., the one reporting itself as "ready to move").
4. *Math interruption.* Participants were asked at random intervals to complete a simple math problem. While this task is somewhat artificial, as platoon commanders are not typically asked to stop and complete math problems in the middle of a heated battle, its cognitive demands are representative of the constant urgent requests they do receive under such circumstances, in which they may be asked to suddenly turn their attention away from the current tasks to address a specific urgent need requiring their military problem solving skills. This task was designed to substantially disrupt their attention on the multiple tasks above. The math task interferes with rehearsal of information kept in working memory (such as counts of civilians, and overwatch status) more so than interruptions from other types of radio messages. In this task, the PDA would beep to alert them that they had a problem awaiting their solution. The participant had to acknowledge the

alert, then the problem would appear on the PDA. Subjects had to solve it, then submit the solution. Performance on the math interruption task was measured in terms of the participant's speed in responding with the answer to the math problem, and the accuracy of the answer.

5. *Mission situation awareness (SA).* During the high task load conditions, participants additionally received "mission messages" pertaining to their current location, the status of various teams and personnel, the overall situation, and their surroundings. These messages were all low priority messages. Performance on this task was measured by a three-question test administered at the end of each trial; participants were asked about the content of the messages that they received.

The high priority tasks in this scenario were: maintain radio counts, maneuver monitoring, and the math interruption task. The low priority tasks were navigation and mission situation awareness. When the EEG and ECG showed that the participant's workload was high, then the Communication Scheduler diverted all radio messages associated with mission situation awareness to the PDA as text messages (there were no messages associated with the navigation task.)

12.4.1.3 Experimental Design We used a two-by-two within subjects design.

Independent variables. There were two independent variables:

- **Mitigation** (mitigated, unmitigated). "Mitigated" means that the Communications Scheduler was available but not necessarily "on." It would only turn on when the EEG and ECG sensors indicated that the participant's cognitive workload was high.
- **Task load** (high, low). Task load was varied by changing the pace at which tasks had to be completed, and the number of simultaneous tasks carried out. In this experiment, subjects in the low task load condition received on average 3.8 messages per minute, and were asked to carry out four tasks simultaneously (all but "mission situation awareness"). In the high

task load condition, participants received on average 8.7 messages per minute and were asked to carry out all five tasks simultaneously.

Note that we have used *task load*, rather than *work load*, as an independent variable. This is because one cannot directly manipulate the participant's cognitive workload. One can only manipulate the task load (e.g., number, complexity, and pace of tasks with the expectation that it will impact cognitive workload. We used our previous experience to set the high and low task loads to levels which we anticipated would result in high and low workloads. A high workload is one that typically degrades task performance. Additionally, in our analysis, we experimentally confirmed that the task loads resulted in the predicted cognitive work loads.

Participants. There were eight male volunteers who participated in this evaluation. They ranged in age from 21 to 42 years of age with an average age of 29.5 years.

Procedure. After signing a consent form, participants put on the equipment: backpack, EEG hat, ECG patch, radio (clipped to shoulder), and hand-held PDA. They received information through the radio and PDA, and they responded through the radio. Participants were then were trained in the field. They first practiced each of the five tasks independently and then all tasks together, all without the Communications Scheduler. They trained until they reached an acceptable level of performance (usually two to three "runs" throughout the field). Next, the cognitive state classifier had to be calibrated for each participant, as each person exhibits workload differently. First, the subject was put in a low task load condition, and the system recorded the EEG and ECG patterns he or she produced. The same was done for the high task load condition.

After the training period, the participants completed four trials as shown in Table 12.1 for every combination of mitigation and task load: mitigation off, task load low; mitigation on, task load high, etc. In the low task load conditions, participants were asked to carry out all but the Mission SA task simultaneously, and they received messages

Table 12.1 Presentation Order of Mitigation in Experimental Trials

PARTICIPANT:	P1	P2	P3	P4	P5	P6	P7	P8
High task load	U(nmitigated)	U	M	M	U	M	U	M
Low task load	U(nmitigated)	U	M	M	U	M	U	M
High task load	M(itigated)	M	U	U	M	U	M	U
Low task load	M(itigated)	M	U	U	M	U	M	U

Table 12.2 Data Collected for the Experiment

TASK	DATA COLLECTED	RESULTS REPORTED	FREQUENCY COLLECTED
Maintain counts	Counts of enemies, friendlies, and civilians	% correct	Multiple per trial
Maneuver monitoring	Number of times participants correctly sent the appropriate squad forward	% correct	Multiple per trial
Interruption task	Response time to interruption alert	Time	Multiple per trial
Interruption task	Number of correct answers to math problem	% correct	Multiple per trial
Mission SA	Number of correct answers to 3-question questionnaire	% correct	Once per high task load trial

at an average of 3.8 per minute. In all conditions, participants were interrupted with a math task twice per minute. In the high task load condition, participants were asked to carry out all five tasks simultaneously, and they received 8.7 messages a minute. In all conditions participants were interrupted twice per minute with math interruption tasks. Performance data are collected for all but the navigation task, as summarized in Table 12.2.

After each trial, subjects were asked to fill out a NASA workload TLX survey. Subjects completing the high task load condition additionally filled out a questionnaire to assess their mission situation awareness. Subjects took a postexperimental questionnaire after completing all trials to gather their opinions on usefulness and challenges in using it.

12.4.1.4 Results We found that the Communications Scheduler did significantly increase performance in three of the tasks: maintain radio counts, mission monitoring, and the math interruption task. However, it decreased performance in the mission situation awareness task. The mission messages for this last task were considered low priority, and therefore were diverted to the PDA during high workload times. However, subjects did not always have time or inclination to review these messages after the high workload period was over. Thus,

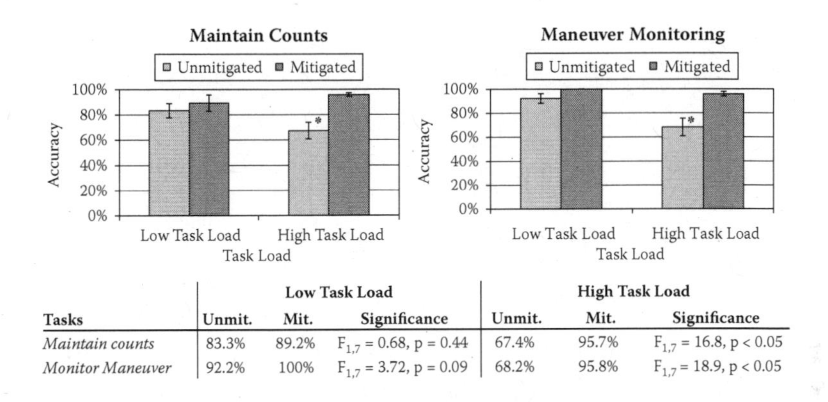

	Low Task Load			High Task Load		
Tasks	Unmit.	Mit.	Significance	Unmit.	Mit.	Significance
Maintain counts	83.3%	89.2%	$F_{1,7} = 0.68$, $p = 0.44$	67.4%	95.7%	$F_{1,7} = 16.8$, $p < 0.05$
Monitor Maneuver	92.2%	100%	$F_{1,7} = 3.72$, $p = 0.09$	68.2%	95.8%	$F_{1,7} = 18.9$, $p < 0.05$

Figure 12.5 Accuracy for the maintain counts and maneuver monitoring tasks. Stars indicate significant differences.

minimizing interruptions has both cost and benefits. If the soldier is not interrupted by low-priority radio messages during especially busy times, the benefit is that he or she can perform more effectively on the high-priority tasks. However, the cost is a loss of situation awareness and effectiveness in the low priority tasks.

The analyses are summarized below. The mitigation produced a statistically significant (alpha < .05) increase in participants' accuracy in maintaining counts and monitoring maneuvers in the high task load condition (see Figure 12.5). In the low task load condition, the mitigation produced no significant change in the participant's performance. This is consistent with the hypothesis that the benefits of the mitigation would be realized primarily during high task load conditions.

Furthermore, when participants were interrupted with a math problem, they were able respond much more quickly in the low task load condition than the high task load condition, indicating the availability of more cognitive resources to attend to the interruption in the low task load condition (see Figure 12.6). In the high task load condition, where it was expected to see the benefits of mitigation, the response time was faster under mitigation by almost 5 seconds. The difference approached significance, however, due to data logging issues; only four of eight participants' data were recorded for the low task load condition and seven of eight participants' data were used in the analysis of the high task load condition. The accuracy of their responses was virtually unchanged in all conditions.

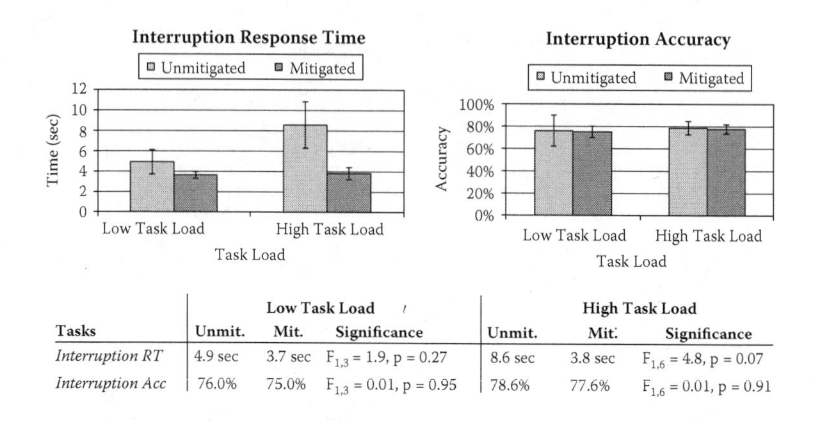

	Low Task Load			High Task Load		
Tasks	**Unmit.**	**Mit.**	**Significance**	**Unmit.**	**Mit.**	**Significance**
Interruption RT	4.9 sec	3.7 sec	$F_{1,3} = 1.9, p = 0.27$	8.6 sec	3.8 sec	$F_{1,6} = 4.8, p = 0.07$
Interruption Acc	76.0%	75.0%	$F_{1,3} = 0.01, p = 0.95$	78.6%	77.6%	$F_{1,6} = 0.01, p = 0.91$

Figure 12.6 Interruption response time and accuracy.

Finally, in the unmitigated, high task load trials, participants' mission SA accuracy was 30% (see Figure 12.7). Participants in the mitigated, high task load scenario were unable to answer any of the questions, since they had not had time to review low priority messages. This effect was statistically significant.

In summary, the Communications Scheduler successfully improved performance of the high-priority tasks during high workload periods. In addition, participants had more attentional resources available with which to react to interruptions, which may be critical if those interruptions are life-threatening emergencies. However, participants' situation awareness of low-priority messages suffered.

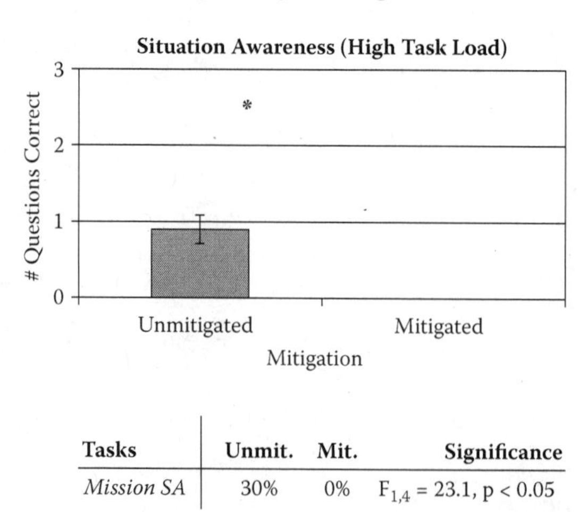

Tasks	**Unmit.**	**Mit.**	**Significance**
Mission SA	30%	0%	$F_{1,4} = 23.1, p < 0.05$

Figure 12.7 Mission SA average number of questions correct. Stars indicate significant differences.

12.4.2 Experiment B: Tactile Navigation Cueing System

The first experiment examined the costs and benefits of minimizing interruptions in high workload situations. This experiment examines a different approach to assisting the soldier in a high workload situation, one in which the computer takes over the planning for one task (navigation) in order to reduce the soldier's burden, but in doing so must also interrupt the soldier with directives. Does such a system ultimately help or hinder the user? This experiment evaluates the costs and benefits of interruptions from the Tactile Navigation Cueing System as it guided soldiers through a mine field.

We tested the hypotheses that:

1. The use of the TNCS would enhance performance on the navigation task during high task load conditions
2. The user of the TNCS would lower performance on some tasks if misapplied during low task load conditions

12.4.2.1 Operational Scenario The setup in this experiment was very similar to the first. Evaluations were carried out in the same field with grass and tree cover. The participant played the role of a platoon leader (PL) whose call sign was Red-6. The Platoon leader managed three squads while reporting to his or her company commander (CO). The commanders and squads were "simulated" as radio messages. However, there were several differences. First, the route in the navigation task was much more complex, and they had to watch out for both mines and enemy cameras. Second, participants were given four tasks to perform simultaneously, rather than five. We used only four tasks because of the increased complexity of the navigation task.

12.4.2.2 Tasks Participants were asked to perform multiple tasks simultaneously, as in the previous experiment.

Navigation. Participants navigated along a complex, unfamiliar, unmarked route. They were given a paper map with landmarks, and possible mine locations. They were asked to navigate to an objective (a goal location) without triggering a mine or being detected by security cameras placed throughout the field. Performance was measured in this task by the time it took to reach the final destination. Participants received a 30-second penalty each time they tripped a mine or were spotted by a security camera.

Maintain radio counts. This task was identical to that in the previous experiment. Performance was measured for this task in terms of the number of correct totals for civilians, enemy soldiers, and friendly soldiers.

Maneuver monitoring. This task was the same as in the previous experiment. Performance on this task was measured by the participant's accuracy in sending the correct team forward (e.g., the one reporting itself as "ready to move").

Math interruption. This task was the same as in the previous experiment. Performance on this task was measured by response time (to acknowledge the interruption), solution time, and the accuracy of the answer generated.

In this experiment, the navigation task was the most important one, particularly avoiding mines and cameras.

12.4.2.3 Experimental Design This was a two-by-two within subjects design.

Independent variables. There were two independent variables:

- **Mitigation** (on, off)
- **Task load** (high, low). In low task load condition, participants received 3.1 radio messages per minute, the route followed was relatively short, and there were only one or two mines and no enemy cameras to avoid. In the high task load condition, participants received 6.3 messages per minute, the route was longer, and they had to avoid on average five mines and two enemy cameras. When task load was high, the TNCS turned on, and the tactor belt started to give directions. No radio messages were diverted to the PDA in this experiment.

Participants. Six male participants volunteered to participate in this evaluation. They ranged in age from 21 to 42 years of age with an average age of 29.5 years.

Table 12.3 Data Collected for the Second Experiment

TASK	DATA COLLECTED	RESULTS REPORTED	FREQUENCY COLLECTED
Navigation	Composite run time (run time + 30 sec penalties)	Time	Once per trial
Maintain counts	Counts of enemies, friendlies, and civilians	% correct	Multiple per trial
Maneuver monitoring	Number of times participants correctly sent the appropriate squad forward	% correct	Multiple per trial
Interruption task	Response time to interruption alert	Time	Multiple per trial
Interruption task	Number of correct answers to math problem	% correct	Multiple per trial

Procedure. All training and trials were run in a large grassy field surrounded by light forest. As in the previous experiment, subjects put on the equipment. Additionally, they were given a map showing their route and possible locations of mines. They trained on all tasks until they reached an acceptable level of performance, without the assistance of the TNCS. Next, the cognitive state classifier was calibrated for each participant. After the training period, the participants completed four trials in a fixed order under, as shown in Table 12.3 (i.e., the same order as in the first experiment). In each trial, subjects had to carry out all four tasks simultaneously while avoiding mines and enemy cameras. The route, mine placement, and camera placement were different in each condition to avoid any learning effects. Participants were interrupted with a math task every 52 seconds on average in both high and low task load conditions. After each trial, subjects were asked to fill out a NASA workload TLX survey. Subjects took a post experimental questionnaire after completing all trials to gather their opinions on usefulness, and challenges in using it. The entire data collection averaged approximately 17 minutes per subject, with a range of 11 to 25 minutes. The data collected for all four tasks is summarized in Table 12.3.

12.4.2.4 Results The TNCS was successful in improving the participants' performance both at navigating the mine field and at secondary tasks, such as accurately completing math problems, without degrading

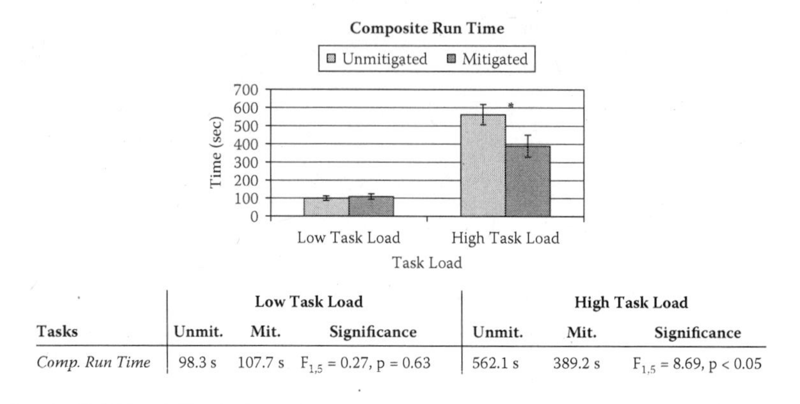

Tasks	Low Task Load			High Task Load		
	Unmit.	Mit.	Significance	Unmit.	Mit.	Significance
Comp. Run Time	98.3 s	107.7 s	$F_{1,5} = 0.27, p = 0.63$	562.1 s	389.2 s	$F_{1,5} = 8.69, p < 0.05$

Figure 12.8 Composite run time. Stars indicate significant differences.

performance on other tasks. We feel that this is because some of the participants' attention resources were freed up when the TNCS took over the navigation task, and in this situation the resources gained outweighed the attention resources lost in responding to the tactor belt's interruptions.

In the high workload condition, the composite run time (e.g., total time to run the course with penalties for tripped mines or detection by enemy surveillance cameras) was significantly shorter when the TNCS was used (see Figure 12.8). Under unmitigated conditions, all six participants triggered mines at least four and up to seven times. With mitigation from the TNCS, however, five of the six participants avoided the mines altogether. Clearly in high task load conditions, participants were able to navigate to the objective more quickly and more safely. These improvements occurred without degrading performance on any of the secondary tasks; in the high workload condition, participants performed equally well on the maneuver monitoring and maintaining counts, regardless of whether the TNCS mitigation was available or not (see Figure 12.9). In fact, the TNCS improved performance on the math interruption task (see Figure 12.10). Again, this confirms that when the TNCS took the burden of the navigation task from the participant, it freed up more cognitive resources to devote to other tasks, such as the math interruption.

However, in low task load conditions, the results were mixed; when using the TNCS mitigation, participants experienced a statistically significant *increase* in accuracy in the math task, but experienced a statistically significant *decrease* in the percentage of times they reported

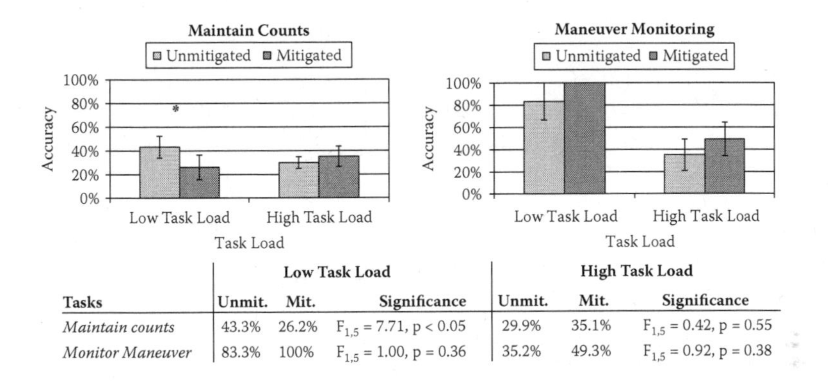

	Low Task Load			High Task Load		
Tasks	**Unmit.**	**Mit.**	**Significance**	**Unmit.**	**Mit.**	**Significance**
Maintain counts	43.3%	26.2%	$F_{1,5} = 7.71, p < 0.05$	29.9%	35.1%	$F_{1,5} = 0.42, p = 0.55$
Monitor Maneuver	83.3%	100%	$F_{1,5} = 1.00, p = 0.36$	35.2%	49.3%	$F_{1,5} = 0.92, p = 0.38$

Figure 12.9 Accuracy for the maintain counts and mission monitoring tasks. Stars indicate significant differences.

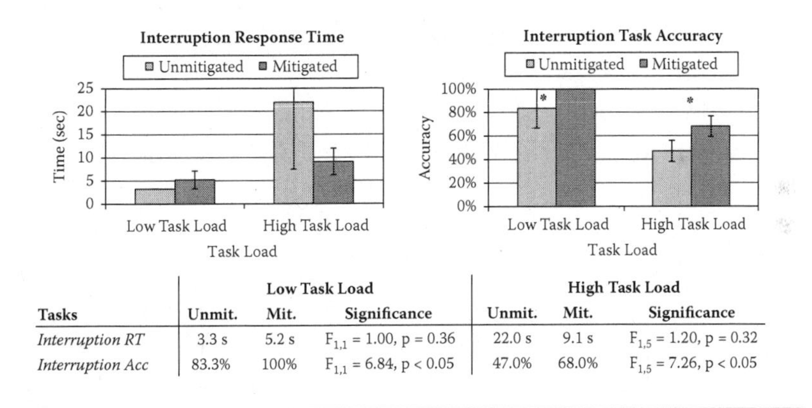

	Low Task Load			High Task Load		
Tasks	**Unmit.**	**Mit.**	**Significance**	**Unmit.**	**Mit.**	**Significance**
Interruption RT	3.3 s	5.2 s	$F_{1,1} = 1.00, p = 0.36$	22.0 s	9.1 s	$F_{1,5} = 1.20, p = 0.32$
Interruption Acc	83.3%	100%	$F_{1,1} = 6.84, p < 0.05$	47.0%	68.0%	$F_{1,5} = 7.26, p < 0.05$

Figure 12.10 Interruption response time and accuracy. Stars indicate significant differences.

correct counts (see Figure 12.9). This is consistent with the hypothesis that the mitigation may result in some performance decrements if applied when it is not needed. This may be because the buzzing from the tractor belt proved to be a significant distraction.

Additionally, we suspect there may have been other costs associated with using the TNCS; however, they did not happen to be costs which we measured in this experiment. For example, we strongly suspect that the TNCS may result in reduced situation awareness; one participant became so focused on the signals from the tactor belt that he forgot to look around him—and he ran into a low-hanging tree branch. Additionally, use of the TNCS may also interfere with learning navigation and terrain information, because the user is passive in the navigation task.

In summary, when users were very overloaded with tasks, the Tactile Navigation Cueing System resulted in performance improvements in almost all tasks by relieving participants of the cognitively challenging task of navigating through an unfamiliar area. However, when the users' tasks were not so challenging, use of the Tactile Navigation Cueing System may impact situation awareness and learning in undesirable ways which warrant future investigation.

12.5 Conclusion

To live and work in the modern world is to be interrupted. Interruptions are typically thought of as undesirable, however, they have both negative and positive consequences. Similarly, minimizing interruptions can also have both positive and negative consequences as these experiments have demonstrated. We examined two adaptive systems, the Communications Scheduler and the Tactile Navigation Cueing System, aimed respectively at assisting mobile dismounted field soldiers during crisis times by (1) minimizing interruptions and (2) taking over a task but interrupting with guidance on that task. We found that minimizing interruptions at critical times enables decision makers to maintain performance on the most critical tasks, but degrades situation awareness on issues that may be currently less pressing, but equally important. We found that when the system took over a task and then interrupted with guidance on that task, it improved users' performance in the short term on almost all remaining tasks by freeing cognitive resources, but it may also have degraded situation awareness overall, and inhibited learning pertaining to the task that was transferred to the computer. Designers of intelligent systems need to be aware of both the *benefits and costs* of systems that interrupt or manage interruptions.

These experiments also provide insights into interactions between humans; there are practical reasons underlying the social etiquette guidelines which surround interruptions. When a speaker interrupts or fails to interrupt, it may have competing consequences on the hearer's effectiveness and situation awareness, and these consequences must be kept in the proper balance. Furthermore, understanding when to interrupt and when to hold back is a complex, highly situation-dependant judgment requiring constant monitoring and attention.

By better understanding the principles which enable humans to live and work together effectively, we may better design computers that can work effectively with people.

Acknowledgments

This research was supported by a contract with DARPA and the U.S. Army Natick Soldier Center, DAAD16-03-C-0054, for which CDR Dylan Schmorrow served as the program manager of the DARPA Improving Warfighter Information Intake Under Stress (AugCog) program and Henry Girolamo was the DARPA agent. The opinions expressed herein are those of the authors and do not necessarily reflect the views of DARPA or the U.S. Army Natick Soldier Center. Additionally, the authors would like to acknowledge the efforts of the Honeywell research and development team that supported the development of the prototypes and the running of the experiments. We would like to make special mention of the contributions of Jim Carciofini, Janet Creaser, Trent Reusser, and Jeff Rye to the work.

References

Adamczyk, P.D., Iqbal, S.T., and Bailey, B.P. 2005. A method, system, and tools for intelligent interruption management. *TAMODIA*, Gdansk, Poland, September 26–27, New York: ACM Press.

Brown, P. and Levinson, S. 1987. *Politeness: Some Universals in Language Usage.* Cambridge: Cambridge University Press.

Chen, D., Hart, J., and Vertegaal, R. 2007. Towards a physiological model of user interruptibility. *INTERACT 2007, Part I.* eds. C. Baranauska et al., 439–451.

Chen, D. and Vertegaal, R. 2004. Using mental load for managing interruptions in physiologically attentive user interfaces. *Proceedings of CHI 2004*, 1513–1516.

Dorneich, M.C., Mathan, S., Ververs, P.M., and Whitlow, S.D. 2007. An evaluation of real-time cognitive state classification in a harsh operational environment. *Proceedings of the Human Factors and Ergonomics Society Conference*, Baltimore, MD, October 1–5.

Dorneich, M.C., Whitlow, S.D., Mathan, S., Ververs, P.M., Erdogmus, D., Pavel, M., Adami, A., Pavel, M., and Lan, T. 2007. Supporting real-time cognitive state classification on a mobile participant. *Journal of Cognitive Engineering and Decision Making* 1(3): 240–270.

Dorneich, M., Whitlow, S., Ververs, P.M., Carciofini, J., and Creaser, J. (2004). Closing the loop of an adaptive system with cognitive state. *Proceedings of the Human Factors and Ergonomics Society Conference 2004*, New Orleans, LA, September 20–24, 2004, pp. 590–594.

Dorneich, M.C., Ververs, P.M., Mathan, S., Whitlow, S.D., Carciofini, J., and Reusser, T. 2006. Neuro-physiologically-driven adaptive automation to improve decision making under stress. *Proceedings of the Human Factors and Ergonomics Society Conference 2006.* San Francisco, CA.

Duchowski, A.T. 2003. *Eye Tracking Methodology: Theory and Practice.* London: Springer-Verlag.

Erdogmus, D., Adami, A., Pavel, M., Lan, T., Mathan, S., Whitlow, S., and Dorneich, M. (2005). Cognitive state estimation based on EEG for augmented cognition, *2nd IEEE EMBS International Conference on Neural Engineering*, Arlington VA, March 16–19, 2005.

Feigh, K. and Dorneich, M.C. (in preparation). A Taxonomy of Adaptive Systems.

Fogarty, J., Hudson, S., and Lai, J. 2004. Examining the robustness of sensor-based statistical models of human interruptibility. *Proceedings of CHI 2004*, 207–214.

Garavan, H., Ross, T.J., Li, S.-J., and Stein, E.A. 2000. A parametric manipulation of central executive functioning using fMRI. *Cerebral Cortex* 10: 585–592.

Gevins, A. and Smith, M. 2000. Neurophysiological measures of working memory and individual differences in cognitive ability and cognitive style. *Cerebral Cortex* 10(9): 829–39.

Gillie, T. and Broadbent, D.E. 1989. What makes interruptions disruptive? A study of length, similarity, and complexity. *Psychological Research* 50: 243–250.

Hancock, P.A. and Chignell, M.H. 1987. Adaptive control in human-machine systems. In *Human Factors Psychology*, ed. P.A. Hancock, 305–345. North Holland: Elsevier.

Horvitz, E. 1999. Principles of mixed-initiative user interfaces. *Proceedings of CHI 1999*,159–166.

Hudson, S.E., Fogarty, J., Atkeson, C.G., Avrahami, D., Forlizzi, J., Kiesler, S., Lee, J.C., and Yang, J. 2003. Predicting human interruptibility with sensors: A wizard of Oz feasibility study. *Proceedings of CHI 2003*, 257–264.

Iqbal, S.T., Adamczyk, P.D., Zheng, X.S., and Bailey, B.P. 2005 Towards an index of opportunity: Understanding changes in mental workload during task execution. *Proceedings of CHI 2005*, 311–320.

Makeig, S. and Jung, T-P. 1995. Changes in alertness are a principal component of variance in the EEG spectrum, *NeuroReport* 7(1): 213–216.

Mathan, S., Whitlow, S., Dorneich, M., and Ververs, P., and Davis, G. 2007. Neurophysiological estimation of interruptibility: Demonstrating feasibility in a field context. In *Proceedings of the 4th International Conference of the Augmented Cognition Society*. October 2007, Baltimore, MD.

Mathan, S., Dorneich, M., and Whitlow, S. 2005. Automation etiquette in the augmented cognition context. *11th International Conference on Human-Computer Interaction (Augmented Cognition International)*, Las Vegas, July 22–27, 2005.

McFarlane, D.C. and Latorella, K.A. 2002. The scope and importance of human interruption in human-computer interaction design. *Human-Computer Interaction* 17: 1–61.

Parasuraman, R. and Miller, C. 2004. Trust and etiquette in high-criticality automated systems. *Communications of the Association for Computing Machinery* 47(4): 51–55.

Pope, A.T., Bogart, E.H., and Bartolome, D.S., 1995. Biocybernetic system evaluates indices of operator engagement in automated task. *Biological Psychology* 40: 187–195.

Powers, R. (2010). United States Army: Chain of Command (Organization). About.com Guide, Available online at: http://usmilitary.about.com/od/army/l/blchancommand.htm?p=1.

Prinzel, L.J., Freeman, F.G., Scerbo, M.W., Mikulka, P.J., and Pope, A.T. 2000. A closed-loop system for examining psychophysiological measures for adaptive task allocation. *International Journal of Aviation Psychology* 10(4): 393–410.

Scerbo, M.W. 1996. Theoretical perspectives on adaptive automation. In *Automation and human Performance: Theory and Applications*, eds. R. Parasuraman and M. Mouloua, 37–63, Mahwah, NJ: Lawrence Erlbaum Associates.

Shell, J., Vertegaal, R., and Skaburskis, A. 2003. EyePliances: Attention-seeking devices that respond to visual attention. *Ext. Abstracts CHI '03*. 770–771, New York, NY: ACM Press.

Spira, J.B. 2005. The high cost of interruptions. *KMWorld* 14(1): 1–32. London: www.ovum.com.

Thorpe, S., Fize, D., and Marlot, C. 1996. Speed of processing in the human visual system. *Nature* 381: 520–522.

Wickens, C.D. and Hollands, J. 2000. *Engineering Psychology and Human Performance*, 3rd edition. Upper Saddle River, NJ: Prentice Hall.

Zeldes, N., Sward, D., and Louchheim 2007. Infomania: Why we can't afford to ignore it any longer. *First Monday* 12: 6–8.

PART VI

The Future: Polite and Rude Computers as Agents of Social Change

13

Etiquette-Based Sociotechnical Design

Brian Whitworth and Tong Liu

Contents

13.1 Introduction

This chapter argues that etiquette applies not just to the people who *use* computers but also to how the technology is *designed*. When technical systems underlie social environments, as in chat, social networks, instant messages, virtual worlds, and online markets, they must support social acts that give synergy, not antisocial acts that destroy it. Politeness, defined as the *free giving of choice to others in social interactions* (Whitworth, 2005), or etiquette, is a critical example. If we don't consider community good when designing sociotechnical systems, we should not be surprised if we don't get it. While etiquette hasn't traditionally been taught in software design courses, we argue that it should be. E-mail spam illustrates what happens when sociotechnical design goes wrong, when it ignores the basic principles of conversation etiquette. Channel e-mail illustrates an etiquette-based design in the illustrative case of e-mail, but the principles proposed are generic to all sociotechnical contexts.

13.1.1 E-Mail Spam

E-mail is perhaps the Internet's primal communication mechanism, being both one of its earliest and most commonly used forms. Yet almost since its inception, it has faced a tide of *spam*—unwanted e-mails from people seeking personal gain. While most of us see spam as a personal inbox problem, the real problem is a community one. Even if all users succeeded in filtering all *their* spam to *their* trashcans, the Internet we share would still have to transmit, process, and store this electronic garbage. The spam your filter "catches" has already wasted Internet resources, and indeed has already been downloaded, processed, and stored by you.

At first, spam seemed more a nuisance than a problem, but in 2003 transmitted spam exceeded nonspam for the first time (Vaughan-Nichols, 2003). So an Internet service provider (ISP) that had one e-mail server for its customers effectively needed to buy another just for the spam. In 2003 over 40% of inbound mail was deleted at the server side by major e-mail providers (Taylor, 2003), though AOL estimated 80% of its incoming 1.5 to 1.9 billion messages a day were filtered as spam (Davidson, 2003). Now spam is the number one unwanted network intrusion, before viruses, and has always been the number one e-mail user complaint.

While *inbox spam* has remained relatively stable, due to spam filter defenses, *transmitted spam* grew from 20% to 40% over 2002/2003 (Weiss, 2003), to 2004 estimates of 60–70% (Boutin, 2004). In 2006 about 86.7% of the 342 billion e-mails sent per year were spam (MAAWG, 2006), and 2007 estimates are as high as 92% (Metrics, 2006). Since transmitted spam consumes processing, bandwidth, and storage whether users see it or not, this is the problem as spam rates increase. While, thanks to filters, many users now find spam a tolerable inbox discomfort, the community problem is growing. Image spam now bypasses text filters, botnets now harvest Web site e-mails, and spammers now use real user e-mails as "zombie" spam sources. Historically, transmitted spam is an e-mail problem that has never stopped growing, so it is a problem that won't be going away anytime soon. Our 2004 prediction that within a decade spam transmission rates will rise above 95% unless something changes seems to be coming true all too soon (Whitworth and Whitworth, 2004).

The worldwide costs of spam to people and machines are staggering, and probably underrated. A 2004 estimate of $1934 per employee year did not include IT staff costs, or hardware, software, and bandwidth wasted by spam (Nucleus Research, 2004). A 2003 estimate of lost productivity for U.S. companies was $10 billion (Bekker, 2003), with a 2005 estimate at $50 billion globally and rising (Ferris Research, 2005).

Why has the human community created a technically advanced communication system where most of the messages are created by one computer then deleted by another shortly after, "untouched by human eye?" An alien viewing our e-mail system would suppose its main function was to transmit messages from one machine to another, rather than from one person to another. The system was designed when the social Internet was still just a dream, to efficiently transmit information not meaning, but today we should know better.

It has been argued that spam is an old social problem in new technology clothes (Whitworth and Whitworth, 2004), essentially an electronic "tragedy of the commons" (Hardin, 1968). In the latter example, a village destroys its commons when everyone seeks individual gain and ignores public good. The common communication "field" we call e-mail is becoming a semantic wasteland in the same way. ISPs operating on limited budgets pay for the Internet bandwidth that spammers consume. Their choices are to let paying customers put up with slower Internet access, to absorb the cost of increasing capacity to pay for the spam, or to raise customer rates. That the spammers who cause the problem pay nothing for its solution is, however, socially unsustainable.

This chapter argues that *sociotechnical problems,* like spam, are technical manifestations of social problems. Hence, they need technology enabled social solutions, i.e., sociotechnical solutions. This chapter is not about etiquette for e-mail users, which is covered elsewhere (see http://www.e-mailreplies.com/), but about reducing "rude by design" software (Cooper, 1999). E-mail, spam, and channel e-mail illustrate a sociotechnical system, sociotechnical problem, and sociotechnical solution, respectively.

13.1.2 Technology Responses to Spam

Most current spam responses are technology-based responses that oppose spam by using technology to combat it directly, without considering social issues, e.g., code filters.

13.1.2.1 Get a New E-Mail Anyone who uses the Internet exposes themselves to spam. To estimate how e-mails get onto spam lists, in 2005 we created dozens of new Yahoo, Hotmail, Gmail, and university e-mail accounts, then counted the spam response for these different online uses:

1. *No actions:* Do nothing
2. *Select newsletter options:* Select to receive Yahoo, Hotmail, and Gmail news notices.
3. *Register e-mail to online gambling sites*: www.grouplotto.com, www.freestuffcenter.com, www.findgift.com
4. *Register e-mail to join online groups:* jokes_to_make_u_laugh@yahoogroups.com, thick_angels@yahoogroups.com, Laugh_Loudly@yahoogroups.com
5. *Register e-mail to educational sites:* www.ftponline.com, www.sun.com, ibm.com
6. *Use in online shopping*: www.ecost.com, www.tigerdirect.com, www.walmart.com

Table 13.1 Average Weekly Spam by Online Actions

ACTION	SPAM
No actions	0
Newsletter options	0
Online gambling	7,177
Online groups	1,720
Educational sites	52
Online shopping purchase	46

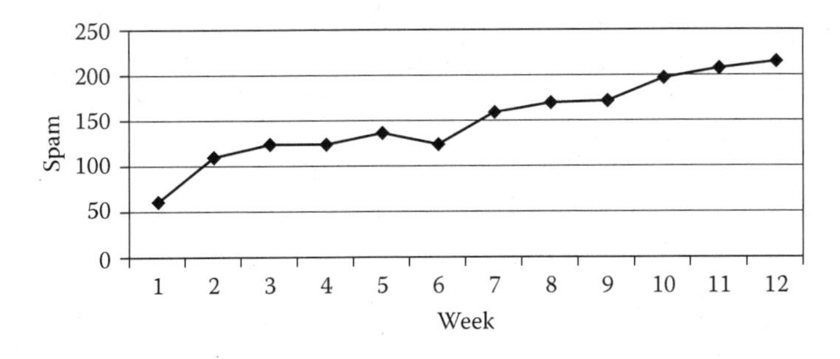

Figure 13.1 Average weekly spam: gambling use.

Table 13.1 shows the total spam for four e-mails over 12 weeks of monitoring. The good news is that while using "risky" online gambling sites gave significant spam, unused new e-mails, even with newsletter options ticked, gave zero spam. E-mails used for online purchases or educational sites gave minimal spam, while online group e-mails gave some spam. The bad news is once an e-mail is listed, spam steadily grows, probably because spammers share lists. In Figure 13.1, while gambling e-mail spam starts at only 50 per week it steadily rises to over 200 per week. For indiscriminate Internet users, the e-mail probably becomes unusable over several years. One solution is to start afresh with a new e-mail, but while this may be financially "free" with Hotmail or Gmail, it is socially expensive as one has to recreate one's social links. Even for cautious e-mail users who only shop online and join groups initially, minimal spam may grow over the years. The long-term effect of spam is to reduce social networks and capital on the Internet.

13.1.2.2 Filters Spam filters, currently the main spam defense, use logic to identify spam content on arrival and put it in the trashcan for later deletion. As machine learning filters improved, with advanced similarity-matching methods and compression techniques (Goodman, Cormack, and Heckerman, 2007), spammers sent more spam to counter their losses, and found new ways to spam (Cranor and LaMacchia, 1998). When machine learning filters identified spam words like "free," spammers wrote "f-r-e-e," or inserted blank HTML comments like "f<!---->ree" which became "free" when rendered. Spammers can now bypass text detection entirely by sending spam images inside random innocuous e-mails impervious to text filters. If providers develop image-matching filters spammers can randomize image content, or if providers block Web image e-mails, spammers can embed images in the message or disassemble it to be reassembled only when the e-mail is rendered (Goodman et al., 2007). There seems to be no end to this arms race as spam "… is almost impossible to define" (Vaughan-Nichols, 2003, p. 42). As the "spam wars" advantage shifts back and forth, the only predictable outcome is that transmitted spam will steadily grow. This war is degrading our common communication system. Filters may even exacerbate the problem, as users isolated behind filter walls are unaware of the rising spam flood.

Filtering before transmission could reduce transmitted spam but has the unintended consequence of hiding filter false positives (real e-mail filtered as spam). E-mail filtered at transmission means the sender is not notified (or spammers could tailor spam to the filter), nor is the receiver notified (as no message is sent). So users could never recover real messages categorized as spam, as we currently occasionally do. E-mails with accidental "spam words" would be filtered, and neither sender nor receiver would know, which affects e-mail trust. An apparent e-mail snub could be one's own ISP filtering the outgoing message in secret. Conversely, one could ignore an e-mail, then claim "the filter took it." The postal ethic that "The mail will get through despite hail, sleet, or snow" creates social confidence, which is critical to any communication system, including e-mail.

13.1.2.3 Lists The lists approach uses lists of who is and is not "a spammer" and checks e-mails against them to identify nonspammers (white list) and spammers (black list). However spammers can easily change identities to avoid black lists, and can "spoof" real users (use them as zombie machines) to get on white ones. Black lists endlessly increase in size, as spammers either create new accounts or spam from a valid account until it is black-listed, then "rinse and repeat" with another account. The long-term sustainability of lists is an issue not just for e-mail but also for all malware, which recently passed the one million known threats mark.

Graylisting uses a combination of black and white lists to reject new e-mails temporarily on the grounds that spammers will move on while real e-mails will try again (Harris, 2003). However, even temporary rejections for 1–4 hours, when messages disappear into an e-mail "limbo," create problems, say, for people awaiting passwords from Web sites.

The administrative effort to create and maintain black/white lists means most individuals don't bother. ISPs maintain such lists, but if one ISP black-lists another, nonspammers in the blocked ISP also have their messages blocked. The logical extension of the list approach is to set up a central spam list; for example, in the Tripoli method (Weinstein, 2003) e-mails need a trusted third party's encrypted "not spam" guarantee to be received. Yet if the "trusted third-parties" are institutional bodies, this raises Juvenal's question: "Quis custodiet ipsos custodies?" ("Who watches the watchers?"). Can major e-mail stakeholders like the Direct Marketing Association, Microsoft, or Yahoo define what is and is not spam? Would they not naturally consider their own "useful services" not spam? If we create a universal e-mail gate, whoever controls it could let themselves in and keep competing others out. A central e-mail "custodian" concentrates power, and history suggests that doing this is a mistake.

13.1.2.4 Challenge Responses Challenge defenses or Human Interaction Proofs (HIPs) check if the sender is human by asking questions supposedly easy for people but not computers; for example, MailBlocks asks users to type in the number shown in a graphic (Mailblocks, 2003). That computers can now answer such questions better than people (Goodman et al., 2007) seems less critical than that most spammers never reply to responses (lest they be spammed). The value of challenges seems to be in the asking itself, not in the question content. Such methods block spam, but e-mail challenges don't save copies, so senders must send e-mail content and any attachments twice. Challenges also increases the psychological cost of sending messages, as senders may take offence at "Are you human?" barriers and not bother.

13.1.3 Social Responses to Spam

Social responses oppose spam with purely social methods aimed at justice. While such methods may use the Internet to find culprits, they do not change computer architectures.

13.1.3.1 Spam the Spammers In simple societies, justice is achieved by individuals seeking "an eye for an eye" revenge (Boyd, 1992). The desire for revenge, to punish wrongdoing whatever the cost to oneself, makes antisocial acts less profitable, as long-term vendetta costs cancel out short-term cheating gains. In Axelrod's prisoner's dilemma tournament the most successful program was TIT-FOR-TAT, which always began by cooperating, but if the other party defected then it did likewise (Axelrod, 1984). In the past, revenge "ethics" may have served a useful social purpose, and Lessig once suggested an Internet bounty on spammers, "like … in the Old West" (Weiss, 2003). The method works for companies who fax annoying unsolicited messages, as users can "bomb" them with return faxes, shutting down their fax machines. However, as Internet spammers usually don't accept replies, spam counterattacks go nowhere (Cranor and LaMacchia, 1998), so revenge methods don't work on the Internet (Held, 1998). Also, revenge has negative side effects—for example, in an online vigilante society, a false Internet rumor could shut down an honest company.

13.1.3.2 Laws Large modern societies bypass vendettas by using "the law," where justice is administered not by individuals but by the state, whose police, courts, and prison or fine sanctions change the social contingencies of antisocial acts. In the jury system the law represents the people to punish antisocial acts. Why not then pass a law against spam on behalf of the online community? This approach does not work for several reasons (Whitworth and deMoor, 2003):

1. Physical laws do not easily transfer online (Burk, 2001; e.g., what is online "trespass?").
2. Online worlds change faster than laws do (e.g., functions like cookies develop faster than they can be assimilated into law).
3. Online architecture is the law (Mitchell, 1995; e.g., programmers can bypass spam laws as if e-mail senders are anonymous, the justice system cannot identify spammers).
4. Laws are limited by jurisdiction. U.S. law applies to U.S. soil, but cyberspace is not within America, because the Internet is global, spam can come from any country.

Laws like the U.S. CAN-SPAM (Controlling the Assault of Non-Solicited Pornography and Marketing) Act fail because the Internet is not under the jurisdiction of any country. A country can "nationalize" their Internet, then claim jurisdiction over it by preventing and monitoring access to "outside" sites or e-mails. China may be moving in that direction by blocking access to controversial entries in sites like Wikipedia or Encyclopaedia Britannica, but such blocking reduces China's information synergy with the rest of the world. For the international Internet to "collapse" into national Internet "tribes" would be a tragedy. Instead of evolving to larger social groups, humanity would be devolving to smaller ones (Diamond, 2005).

On a practical level, legal prosecutions require physical evidence, an accused, and a plaintiff, but spam can begin and end in cyberspace. With easily "spoofed" e-mail sources, and a "plaintiff" that is everyone with an e-mail, what penalties apply when identity is unclear and each individual loses so little? The law still cannot contain *telephone* spam (telemarketing), let alone *computer* spam. Traditional law seems too limited, too slow, and too impotent to deal with the global information challenge of spam. This failure suggests the need for an approach that engages rather than ignores technology.

13.1.3.3 Summary Technology methods alone create an "arms race" between spam and filters, while purely social methods like the law work but are ineffective in cyberspace. Neither social responses (revenge, laws) nor technical responses (filters, lists, challenges) have stemmed the spam tide. Technology responses ignore the social contingencies that create spam, and social responses ignore the realities of software architectures that enable spam. It is time to try a socio-technical approach, which fits technical architectures to social principles.

13.1.4 Sociotechnical Responses

The sociotechnical approach to software design involves two steps:

1. *Analysis.* Analyze what social forms are desirable (e.g., polite conversation).
2. *Design.* Design a technology architecture to support these forms.

Note that enforcing "good" by police-state-style tactics is incompatible with sociotechnical axioms of legitimacy, transparency, freedom, and order (Whitworth and Friedman, 2009). The method explicitly defines a desired social process, here a polite conversation, then translates its requirements into a technical design that supports both social and technical needs.

13.1.4.1 An E-Mail Charge While filters stop spammers directly, social methods reduce spam indirectly by introducing negative consequences for antisocial acts, like prison or fines. This method works for most, and the few immune to social pressure can be hunted down and isolated in prison by crime units. However an anonymous Internet makes this much more difficult; registering every user opens online society to the risk of takeover or hijack, even if all nations could agree to it.

An e-mail charge, applied to everyone, could let the Internet remain decentralized by following the economic principle that people try to reduce their costs, that is, hit spammers in their pockets. In information terms, every transmission could extract a micro-payment, or all senders could compute a time-cost function trivial for all e-mail, so spammers would find the cost excessive (Dwork and Naor, 1993). Such methods essentially suggest redesigning software to increase e-mail transmission costs. However, e-mail charges also reduce overall usage (Kraut et al., 2002), so stopping spammers by slowing the e-mail flow, whether by unneeded charges or pointless calculations, seems like burning down your house to prevent break-ins. Also it is politically difficult to justify introducing a new charge for services we already have. A new Internet "toll" would add no new service above those now available, as e-mail already works without charges.

Finally, making the Internet a field of profit opens it to corruption. If senders paid receivers, and each e-mail transferred money, who would administer the system and set the charge rate? Is an administration charge effectively an e-mail tax? If so, who then will "govern" online e-mail? Spam works because no-charge e-mail is easy, which is also why the Internet itself works. Its decentralization, with no one "in charge," is why the Internet has so far largely resisted corrupt takeover. The social principle of charging for e-mail contradicts the original principle of e-mail success, namely that fast, easy, and free communication benefits everyone. This is the principle of *social synergy* lying behind the success of the Internet itself. A solution is needed that reduces spam but still leaves the Internet advantage intact.

13.1.4.2 Analysis Spam works under the following conditions:

1. *Benefit.* With an online sucker born every minute, whatever the pitch, someone always takes the bait.
2. *Cost.* E-mail is so cheap it costs little more to send a million e-mails than to send one.
3. *Risk.* The problems spammers create for others have no consequences for themselves.
4. *Ability.* Spammers do what they do because technology makes it possible.

If the response percentage is always positive (#1), the extra message cost near zero (#2), the consequences zero (#3) and e-mail tools provide the ability (#4), is not spam inevitable? Filters try to remove condition #1, but it is a losing battle. With a billion plus worldwide e-mails and growing, spammers need only one hundred takers per ten million requests to earn a profit (Weiss, 2003). Even with filters 99% successful, which they aren't, a hit rate as low as 0.001% still makes spam profitable. The predicted end-point is spammers targeting all users, giving a system that technically "communicates" but mainly in messages no one reads.

13.1.4.3 The Social Requirements of Technical Communication While the first condition seems an inevitable part of human nature, and the second a desirable technology advantage, the third and fourth seem just plain wrong. Technology shouldn't let spammers create negative consequences for others at no cost to themselves, as it doesn't happen in face-to-face interactions.

On a technical level, e-mail is an information exchange, but on a human level it is a meaning exchange that has a name—a *conversation*. Face-to-face conversations have *etiquette requirements* (e.g., just walking up to strangers and conversing usually produces a negative response). People expect others to introduce themselves first, so the other can decide to converse or not, depending on who you are. One may even ask a friend, "Can we talk?" if they look busy. Likewise filibustering at a town hall meeting will get you ejected, as people are supposed to let others speak. The social principle that *conversations are mutual* is an agreed etiquette, based on the politeness principle that each gives choices to the other.

Technology changes the social dynamics (e.g., telephones let anyone call anyone), creating the telemarketing problem of unknown sellers calling at dinner times, reducing the social capital and synergy of society. So society implemented telemarketing laws, as it did when the postal system similarly let bulk mail companies send out masses of unwanted "junk" mail. While the architecture has changed from postal to telephone to e-mail, the social response is the same: to assert *the right not to communicate*, to be left alone or to remain silent. This right is framed negatively because laws and sanctions work that way, but can be stated positively: that *communication requires receiver consent*. The underlying social principle is freedom, and the reason societies adopt it is that free societies are more productive (Whitworth, 2005).

Such social concepts are subtle, as viewers don't mind noninvasive media like billboards or subway ads that they can ignore. Likewise, television viewers are free to change channels to avoid advertisements, so it is their choice to watch them or not. In contrast, online pop-up ads that hijack the current window, and the annoying Mr Clippy who hijacked the user's cursor, allow no choice. Similarly e-mail spam comes whether one wants it or not; there is no choice as to even delete a spam message, one must look at it. It is the forcing of communication that is the problem. While spam may not register as strongly as antisocial acts like stealing, murder, or rape, it is in the same category of social acts where one party forces another do something they don't want to do. Spam is not a victimless crime just because its effects are spread across millions. The current spam plague illustrates what happens when the social right to consent to communicate is ignored by a technology design. In contrast, more recent technologies like online chat give more choice (e.g., one can't join a chat unless current users agree).

By this analysis, the current e-mail architecture that lets senders place messages directly into a receiver's inbox is socially wrong because it gives users the right to *communicate unilaterally.* Instead of giving senders all rights, and receivers no rights at all, e-mail should share the right to communicate between sender and receiver (Whitworth and Whitworth, 2004). In file transfer, the opposite problem occurs; receivers have all rights and senders have none, causing the social problems of plagiarism and piracy. In online communication, control should be shared (Duan, Dong, and Gopalan, 2005), as per the following social requirements:

1. All conversations require mutual consent.
2. One has the right to refuse any conversation.
3. Anyone can request to converse with another.
4. Once a conversation is agreed, messages can be exchanged freely, usually in turns.
5. One can exit a conversation at any time.

Channel e-mail is a technical design that meets these requirements by sharing communication control between sender and receiver.

13.1.4.4 Channel E-Mail The channel e-mail protocol supports conversation requirements via a conversation "channel," an entity that sits above the messages sent. Instead of managing messages, channel e-mail users manage channels (which are less numerous, as one channel has many messages). A channel's recency is that of its last message, as in Gmail threading. An open channel grants mutual rights to freely send messages between online parties, as in current e-mail. Only if there is no open channel must one be negotiated, via channel "pings"—small e-mails with permissions. Opening a channel is a separate step from sending a message, like the handshaking of face-to-face conversations and some forms of synchronous communication. The handshaking can be automated, letting users just send their messages while the computers negotiate the permissions.

Instead of the current "send and forget" one-step protocol, channel e-mail has several steps:

1. *Channel request.* A conversation request (A to B).
2. *Channel permission.* A permission to converse (B to A).
3. *Message transmissions.* Conversation messages are transmitted mutually.
4. *Channel closure.* Either party closes the channel.

Messages in step #3 use the channel open permission, so further messages do not need channel requests. Channel control is not just the receiver's right to tediously reject e-mails one by one, but the right to close a channel entirely, including all future messages from that source. Aspects of this approach are already in practice. For example, Hotmail recognizes:

1. *Safe senders:* Senders who are granted a channel to send e-mail.
2. *Blocked senders*: Senders who are blocked from sending e-mail (i.e., a closed channel).

DiffMail handles spam by classifying senders into (Duan et al., 2005):

1. *Regular contacts*: Message header and content are sent (pushed) to the receiver inbox.
2. *Known spammers*: Messages are not delivered.
3. *Unclassified*: These messages must be retrieved (pulled) by receivers.

In channel e-mail, the receiver effectively "pulls" a new message by sending a permission to the sender's "push" request. Users can manage channels by setting them as:

1. *Open.* Always accept messages.
2. *Closed.* Always reject messages.
3. *Undefined.* Ask me each time.

Meta-option defaults for new (unknown) channel requests can be set as:

1. *Always accept:* The equivalent of current standard e-mail (i.e., completely public).
2. *Always reject:* Closed off to all unknown e-mails (i.e., completely private).
3. *Ask me:* User decides each time.

The operation and feasibility of channel e-mail are now considered.

13.1.4.5 Operation Channel e-mail reduces the number of screen lines to manage as it shows channel threads, each containing many messages, in order of the recency of its last message. In Gmail such *threading* keeps same conversation messages together and avoids flipping between "inbox" and "outbox" to figure out who said what last. Channel e-mail has no inbox or outbox, as it threads solely by sender, not topic. If one sender has many e-mail aliases, they can be designated to the same channel. A Gmail sender who changes a message title starts a new thread, but in channel e-mail changing, the topic doesn't change the conversation thread.

In channel e-mail users manage approved sender channels as they manage their friends in Facebook. All unknown senders, including spammers, go into a separate "Channel Requests" category. Only socially näive software would muddle known and unknown sender messages into one inbox that doesn't discriminate friend from stranger.

Channel e-mail users could manage their channels or just send and receive e-mails as usual with the software handling channel requests (i.e., let the handshaking occur in the background). The user setting "Always accept new channel requests" automatically returns a permission to an unknown sender, which they could then use. A new sender without channel e-mail would receive the permission as an e-mail, explaining that they can reply to this e-mail to use the permission. This minor change seems to permit communications from both spammers and nonspammers, but spam would immediately reduce as spammers almost never reply to e-mail lest they get spammed in return. Conversely, for people it would be a natural etiquette to get permission before sending a first-time message.

Channel e-mail gives receivers a choice over who they talk to by "democratizing" list methods, letting users create personal black and white lists based on channels. It does not reject ISP controlled lists, but just lets users make their own choices as well. List maintenance, a problem for centralized lists, occurs easily at the local level as every sent message implies an open channel and every rejected message closes a channel. Users naturally opening and closing channels in normal e-mail use effectively define their local lists. Channel e-mail gives e-mail users a choice they should have had in the first place.

13.1.4.6 Feasibility Challenge e-mail systems already use a three-step send–challenge–resend protocol. The challenge "Captcha" question is a task supposedly easy for people but hard for computers, like to read a blurry word. A sender who satisfies the challenge can resend the e-mail. If the three-step challenge protocol can be implemented, so can the request–permit–send protocol of channel e-mail. Yet many find "Are you really a person?" humanity challenges insulting and dislike them. In contrast, in channel e-mail it is *polite* to ask if one can send someone an e-mail for the first time. Real people know that relationships take work.

Channel requests would include sender name, e-mail, and subject but not message content or attachments. These "pings" can be of minimal size, perhaps involving:

1. *Title* (e.g., "Can I talk to you about <topic>? Press reply to receive my message.")
2. *Channel properties:* Sender e-mail/name/IP address, receiver e-mail/name/IP address, request date/time, accept date/time.

Permissions could use e-mail properties like sender IP address, and request received date/time or even user defined tags. They are dynamic, so closing and reopening a channel creates new permissions. They could even be tags visible in the subject line, so publishing an e-mail tag like "happyvalley99" on a Web site could let readers paste it into an e-mail title to get a direct channel. This is not designed to be secure but interactive: if it is compromised, one can just reset the permission.

Publishing tags could help classify incoming e-mails (e.g., if class students are given a tag to use for all class e-mails, their messages will automatically sort into a channel, rather than into an already overflowing "inbox" which users must organize manually or by setting complex filters). Channel e-mail users could ask unknown e-mails to use designated channels: "Please use one of my public e-mail tags in your title: [mycompany], [myname] or [myhobby]." As well as open and close, users could create, delete, merge, and transfer channels. Design options like group channels and public key channels are beyond what can be outlined here.

13.1.5 Theoretical Evaluation

While a spammed network may be technically efficient (in bytes/second), it is socially inefficient if most of the messages transmitted are spam. *Social efficiency (SE)* can be defined as the proportion of socially useful bytes sent (for a given time period):

$$SE = \frac{\text{Nonspam bytes sent}}{\text{Total bytes sent}}$$

This ratio is the proportion of network resources used to send socially useful messages. For SE = 100% all network resources are sending non-spam messages, while SE = 40% means only 40% of the network capacity transmits useful messages and 60% transmits spam messages deleted by filters. A network is *technically efficient* if it transfers information well, but *socially inefficient* if it mostly sends spam no one reads or wants to read.

Using an average 59 KB e-mail size (Szajna, 1994), an average 12 KB spam size (Willams, 2007), and an estimated request size of 0.25 KB, Figure 13.2 compares the theoretical social efficiency of standard and channel e-mail by *spam rate*—the percentage of all e-mails that are spam. It was done for the "worst case" scenario (for channel e-mail), where *all* senders are new contacts needing channel permissions. As shown, while standard e-mail begins 100% socially efficient, under spam assault its efficiency declines rapidly. In contrast, while channel e-mail begins at less than 100% efficient, it remains relatively stable under spam load.

The size of each spam message also significantly impacts social efficiency, as larger spam messages, such as those with images or attachments, use up more network resources. Figure 13.3 shows how standard e-mail social efficiency degrades rapidly as spam *message size* increases for a spam rate of 80%, again for a worst-case scenario (for channel e-mail) where all messages are sent by new contacts. In contrast, message size only minimally affects the social efficiency of channel e-mail.

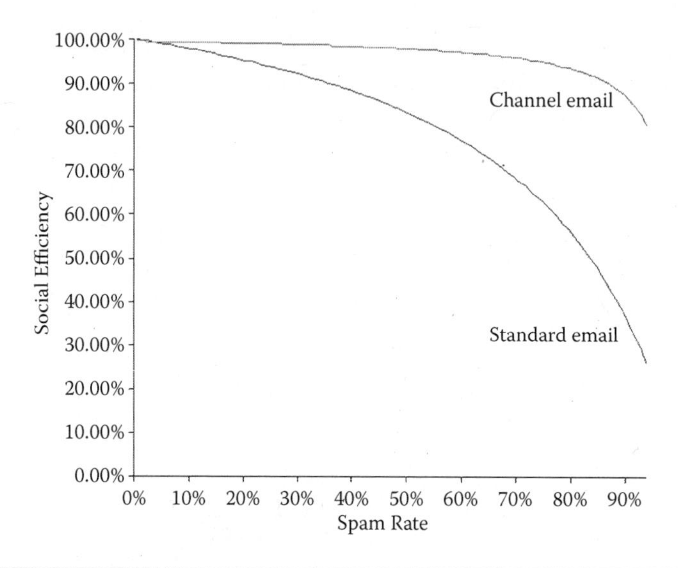

Figure 13.2 Social efficiency by spam rate (theoretical).

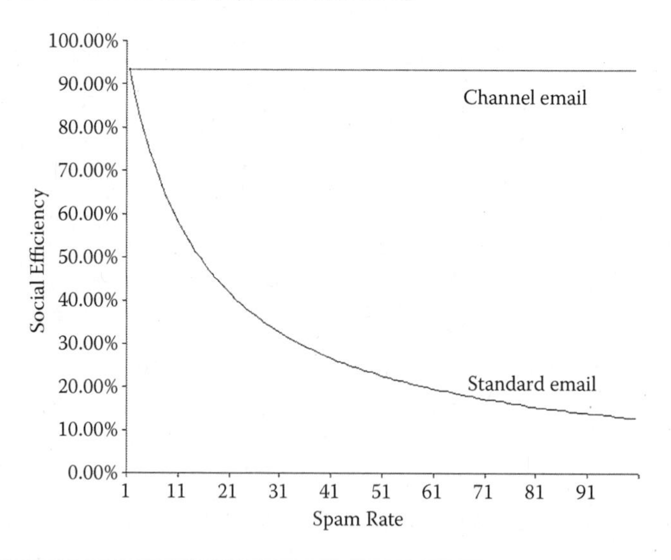

Figure 13.3 Social efficiency by spam message size (theoretical).

13.1.6 Simulation Evaluation

To verify these theoretical outcomes a channel e-mail communication simulation was set up. One computer sent messages to another over a local network isolated from outside influences, including the Internet. In the standard mode messages were just sent, but in the channel mode messages required channel permissions. A third computer

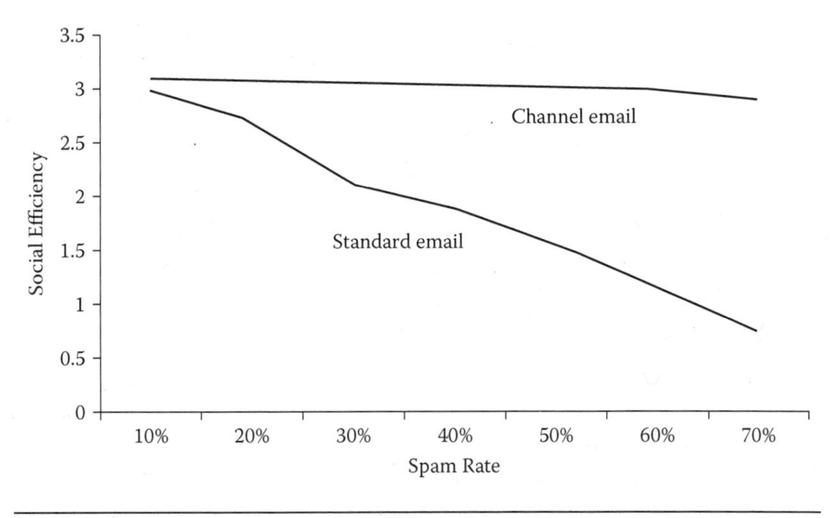

Figure 13.4 Social efficiency by spam rate (actual/simulated).

simulated an outside spam source, using spam message sizes of small (5 KB), medium (10 KB) and large (60 KB), sent at spam rates from 10% to 70% of the nonspam messages. Network social efficiency was estimated by measuring nonspam message *transmission rate* (MB/sec), which was high if nonspam messages arrived quickly and low if they took a long time due to the spam load. This was taken as a valid measure of social efficiency. Figure 13.4 shows how social efficiency declined by spam rate for medium-sized spam messages. While standard e-mail decreases drastically under spam assault, channel e-mail performance is again robust. Increases in message size also drastically affected standard e-mail but had little impact on channel e-mail.

13.1.7 User Evaluation

Finally, whether users would accept channel e-mail was evaluated by creating two matching Web-based e-mail prototypes and comparing their usability:

1. *Standard e-mail:* had an inbox with messages received and an outbox with messages sent.
2. *Channel e-mail:* showed channel requests, sent e-mails, and current channels. Current channels were divided into open (known), closed (spam), and not yet classified areas.

In the channel prototype e-mails from first time senders went into a channel requests area, where users could press buttons to “Accept sender” (move to open channel area) or to “Reject sender” (move to spam area). It could later be moved again to another area (e.g., to unclassified). This study evaluated user interface acceptance rather than network performance, so the prototype did not use a three-step channel protocol, but just sent simple messages. Subjects were in groups of 10, each with an allocated e-mail ID. Their task was to send 17 simple e-mail questions to other participants, like “What is your birthday?” and also to respond to 17 such questions from other participants. A total of 34 valid e-mails had to be sent and replied to per person, which took on average 43.4 minutes. In addition, incoming spam was sent to all participants at rates of 12%, 40%, or 73% of valid e-mails sent. Subjects were divided randomly into two groups, one evaluating the traditional e-mail prototype, and the other the matching channel e-mail prototype. Each group completed the task for each of three spam levels, then completed a questionnaire with four usability dimensions: Understandability, Learnability, Operability, and Perceived Usefulness, based on a validated model (Bevan, 1997). The survey questions were:

1. I would find it easy to get this e-mail system to do what I want to do. (Understandability)
2. To learn to operate this e-mail system would be easy for me. (Learnability)
3. I would find this e-mail system to be flexible to interact with. (Operability)
4. Using this e-mail system in my job would enable me to accomplish tasks more quickly. (Perceived Usefulness)
5. Using this e-mail system would make it easier to do what I want to do. (Perceived Usefulness)

The seven-point response scale was from extremely likely to extremely unlikely.

Figure 13.5 summarizes the results, where the channel e-mail interface rated higher than standard e-mail on both usability and usefulness dimensions. A t-test comparison of the mean response scores for questions 1–5 for standard versus channel interfaces was significant at the 0.01 level, suggesting users clearly preferred channel e-mail.

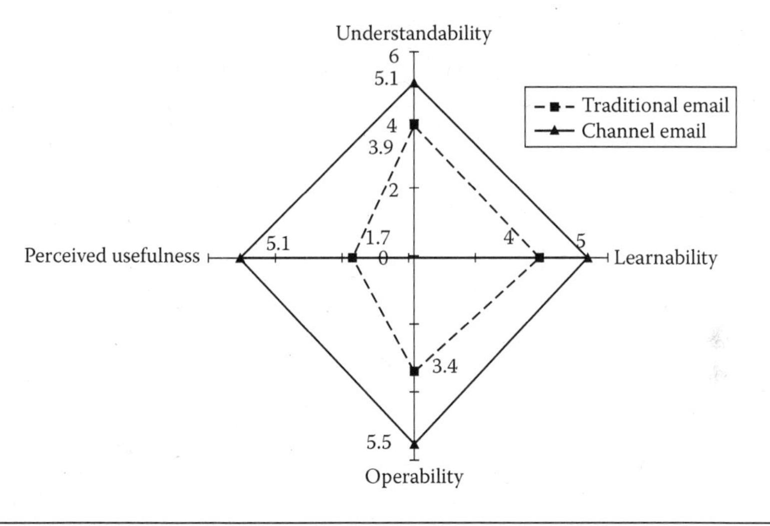

Figure 13.5 A usability comparison of standard versus channel e-mail.

13.1.8 Implementation

To deploy a new system over an existing one it must be backwards compatible; to grow and develop, it must also be useful.

13.1.8.1 Compatibility For a useful new system to catch on it must first survive an introductory period when only a few people in the community actually use it. To survive this phase, it must be backward compatible with existing, more broadly used systems. Channel e-mail has two such cases:

1. *Non-channel sender*: Channel e-mail can treat e-mails from new senders as channel requests, and reply to accepted senders with a channel permission: "Press Reply to send your e-mail by this channel." Since the sender may not be familiar with the channel idea, it would include a "canned" explanation of what a channel is and why spam makes it necessary.
2. *Non-channel receiver*: In this case, the permit ping; the channel that appears as a first time e-mail to someone else, appears as a polite request: "XYZ wants to send you an e-mail on <topic> Just press reply to receive the e-mail." Again, it could explain channel e-mail, and that this is a once-only connect request. The pending message would automatically be sent on receiving a reply.

Users converting to channel e-mail could minimize the impact on their friends by sending out channel invitation e-mails using their address book. In this system it takes effort to manage one's friends, as cell phone users for example do. Once set up, a channel is easy to use, but spammers must now do work to establish a new communication channel.

13.1.8.2 Usefulness Channel e-mail modifies and combines features from *black/white lists* and *challenge e-mail.*

It democratizes black/white lists, as users can personally define who they talk to by opening and closing channels. We naturally know this, so systems like Facebook succeed by giving users tools to manage their friend/not-friend lists. Spammers can bypass ISP controlled black lists of known spammers by creating a new identity, but as individuals usually work by white lists of friends, a spammer with a new identity is still "unknown."

Channel e-mail has no "Are you human?" challenge, but its three-step process challenges the sender to reply to the permission, which spammers almost never do. *That spammers never reply is the perfect spam filter.* If spammers adopt channel e-mail to blend in, transmitted spam would still be reduced as Figure 13.6 shows. Channel e-mail gives a return on network resources for all spam rates over 10%. At an 80% spam rate it saves about 40.8% percent of network resources, so at current rates it would allow an Internet service provider to replace three e-mail servers by two.

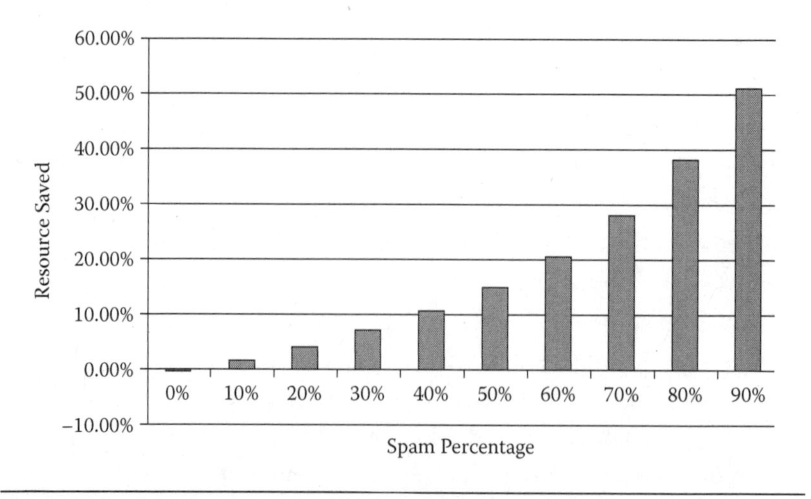

Figure 13.6 Network resources saved by channel e-mail, by spam rate.

13.1.8.3 Organizational Spam Channel e-mail also works with *organizational spam* where both filters and lists fail. While in commercial spam a few people send millions of unwanted messages, organizational spam is many people sending unwanted messages, like: "My daughter needs a piano tutor, can anyone recommend one?" This message was actually sent to everyone in a large organization. While commercial spammers epitomize selfishness, organizational spammers are just ordinary people being inconsiderate. Yet it is as big a problem as commercial spam. When people spam a community list, standard e-mail only lets one unsubscribe, but channel e-mail gives two more options

1. *Return to sender.* While commercial spammers don't receive e-mails, organizational spammers do. The *Return to Sender* button deletes the message and forwards it back to the sender with a "No, thank you" note (Whitworth and Whitworth, 2004). Under channel e-mail, spamming a group list could results in hundreds or thousands of return-to-sender replies.
2. *Close the channel.* Choosing this option automatically creates an e-mail such as "The user has closed this channel. To reopen ..." and any further contacts from that person become new channel requests. This doesn't close the list, just that sender.

Channel e-mail lets the community give feedback to those who abuse its goodwill. Since people are often extremely sensitive to social censure, the social effect on those who routinely send unwanted e-mail messages to community lists could be dramatic.

13.1.9 Discussion

The approach taken here differs markedly from Microsoft's "more of the same" approach, where researchers feel smart filters are "holding the line," and that *we* will defeat *them* in the spam wars (Goodman et al., 2007). Online history doesn't support this view. Indeed, it is predictable that two sides with the same assets, of human cunning and computer power, will inevitably produce an endless spam arms race, and indeed transmitted spam has never dropped. Spam researchers may rejoice that spam wars will be "keeping us busy for a long time to come" (Goodman et al., 2007), but why use our Internet commons as your battleground?

Simple arithmetic suggests that technology alone cannot contain the spam challenge forever; with over 23 million businesses in America alone, each sending just one unsolicited message per year to all users, there are over 63,000 e-mail messages per person per day. Spam potential grows geometrically with the number of users, and with billions online in the future it easily outstrips Moore's law of technology growth.

If e-mail "dies" from spam overload, don't imagine we can just move to other applications as the spam plague already infects them, e.g., SPIM (IM spam) and SPAT (chat spam). Acting like slash-and-burn farmers is not an option. Spam is just the poster child for a genre of antisocial acts threatening online society, including spyware, phishing, spoofing, scams, identity theft, libel, privacy invasions, piracy, plagiarism, electronic harassment, and other Internet "dark side" examples. Spam is a social problem that online humanity must eventually face.

An ideal world might have no spam but in our world, it is a reality. Yet it needn't overwhelm us, as societies have managed antisocial behaviors for thousands of years. By *social evolution* ideals like legitimacy, democracy, transparency, and freedom have emerged (Whitworth and deMoor, 2003). Politeness is one of these, and as such deserves support.

Spammers are the commercial fishing trawlers of the information world, whose huge technology-created nets making previously abundant environments barren. Obviously taking continuously from a physical system without returning anything is not a sustainable process, but it is less obvious that this applies equally to social and informational systems. One option is to introduce fair play rules that limit "catch" sizes. In social systems, the positive path to the same end is to encourage people to give consideration to each other by supporting an etiquette.

Traditional society punishes antisocial acts by sanctions like prison, but the Internet can support social acts by technical design. In "polite computing," code supports beneficial ideals like fairness and democracy; for example, channel e-mail gives senders and receivers equal rights to communicate. Online social rights are defined by what the computer code allows, which is as if physicists could define the laws of physics in physical space. As we create online social environments, the code that creates the anarchy of spam also allows Orwellian control of all online user expression. On this question of how our sociotechnical environment will be built, software designers cannot sit impartially on the sidelines. Let us embed what physical society has learned socially over the centuries, written often in blood, in sociotechnical designs. Principles like freedom, transparency, order, and democracy deserve design support (Whitworth and Friedman, 2009). The role of etiquette and politeness in this is to be a positive force. While *laws* define what people *must* to in legitimate social interactions, *politeness* defines what they *could* do to help each other. While one aims to obstruct "evil," the other aims to direct "good." Sociotechnical software that supports positive social interaction by design is better than trying to chase down and punish negative interactions.

Hence, channel e-mail doesn't try to find or punish spammers. It doesn't target them at all. Its target is the unfair social environment that "grows" spam. Fair rules apply to all equally, so anyone can choose to converse or not, and any e-mail can be rejected, not just spam. The goal is fair communication, not punishment or revenge. A community that is given the tools to discern and choose between social and antisocial acts will choose what helps it survive. As transmitted spam moves steadily to an equilibrium end-point of over 99%, the value of positive sociotechnical design will become increasingly apparent.

Acknowledgments

Thanks to Deepthi Gottimukkala, Srikanth Kollipari, and Vikram Koganti for the spam data, to Victor Bi for the channel e-mail interface, and to Zheng Dai for network simulations.

References

Axelrod, R. (1984). *The Evolution of Cooperation*. New York: Basic Books.

Bekker, S. (2003). Spam to cost U.S. companies $10 billion in 2003. *ENTNews*, October 14.

Bevan, N. (1997). Quality and usability: A new framework. In A. M. E. van Veenendaal, J (Ed.), *Achieving Software Product Quality*. Netherlands: Tutein Nolthenius.

Boutin, P. (2004, April 19). Can e-mail be saved? *Infoworld,* 41–53.

Boyd, R. (1992). The evolution of reciprocity when conditions vary. In A. H. Harcourt and F. B. M. de Waal (Eds.), *Coalitions and Alliances in Humans and Other Animals*. Oxford: Oxford University Press.

Burk, D. L. (2001). Copyrightable functions and patentable speech. *Communications of the ACM*, 44, February(2), 69–75.

Cooper, A. (1999). *The Inmates are Running the Asylum: Why High Tech Products Drive Us Crazy and How to Restore the Sanity.* Indianapolis, IN: SAMS.

Cranor, L. F. and LaMacchia, B. A. (1998). Spam! *Communications of the ACM, 41*(8), 74–83.

Davidson, P. (2003, April 17). Facing dip in subscribers, America Online steps up efforts to block spam. *USA Today,* p. 3B.

Diamond, J. (2005). *Collapse: How Societies Choose to Fail or Succeed.* New York: Viking (Penguin Group).

Duan, Z., Dong, Y., and Gopalan, K. (2005). A differentiated message delivery architecture to control spam. Paper presented at the *11th International Conference on Parallel and Distributed Systems (ICPADS'05).*

Dwork, C. and Naor, M. (Eds.). (1993). *Pricing via Processing or Combatting Junk Mail.* In E. Brickell (Ed.), Advances in Cryptology-Crypto '92, Lecture notes in Computer Science 740 (pp. 139–147), New York, NY: Springer-Verlag.

Ferris Research. (2005). Reducing the $50 billion global spam Bill [Electronic Version]. *Ferris Report, February* from http://www.ferris.com/?page_id=73258.

Goodman, J., Cormack, G. V., and Heckerman, D. (2007). Spam and the ongoing battle for the inbox. *Communications of ACM, 50*(2), 25–33.

Hardin, G. (1968). The tragedy of the commons. *Science, 162*, 1243–1248.

Harris, E. (2003). The next step in the spam control war: Greylisting. From http://projects.puremagic.com/greylisting/whitepaper.html.

Held, G. (1998). Spam the spammer. *International Journal of Network Management, 8*, 69–69.

Kraut, R. E., Shyam, S., Morris, J., Telang, R., Filer, D., and Cronin, M. (2002). Markets for Attention: Will Postage for E-mail Help? Paper presented at the CSCW 02, New Orleans.

MAAWG. (2006, June 2006). MAAWG E-mail Metrics Program, First Quarter 2006 Report. Retrieved April 1, 2007, from http://www.maawg.org/about/FINAL_1Q2006_Metrics_Report.pdf.

Mailblocks. (2003). Mailblocks is the ultimate spam-blocking e-mail service.

Metrics. (2006). 2006: The year spam raised its game; 2007 predictions [Electronic Version]. MessageLabs December from http://www.metrics2.com/blog/2006/12/18/2006_the_year_spam_raised_its_game_2007_prediction.html.

Mitchell, W. J. (1995). *City of Bits Space, Place and the Infobahn*. Cambridge, MA: MIT Press.

Nucleus-Research. (2004). Spam: The Serial ROI Killer [Electronic Version], Research Note E50 from http://www.nucleusresearch.com/.

Szajna, B. (1994). How much is information systems research addressing key practitioner concerns? *Database*, May, 49–59.

Taylor, C. (2003). Spam's Big Bang. *Time*, June 16, 50–53.

Vaughan-Nichols, S. J. (2003). Saving private E-mail. *IEEE Spectrum, 40*(8), 40–44.

Weinstein, L. (2003). Inside risks: Spam wars. *Communications of the ACM, 46*(8), 136.

Weiss, A. (2003). Ending spam's free ride. *netWorker, 7*(2), 18–24.

Whitworth, B. and Friedman, R. (2009). Reinventing academic publishing online, Part II: A Socio-technical vision. *First Monday, 14*(9).

Whitworth, B. (2005). Polite computing. *Behaviour and Information Technology, 24*(5, September, http://brianwhitworth.com/polite05.pdf), 353–363.

Whitworth, B. and deMoor, A. (2003). Legitimate by design: Towards trusted virtual community environments. *Behaviour and Information Technology, 22*(1), 31–51.

Whitworth, B. and Whitworth, E. (2004). Reducing spam by closing the social-technical gap. *Computer* (October), 38–45.

Willams, I. (2007). Image spam doubles average file size. Retrieved October 10, 2007, from http://www.vnunet.com/articles/print/2186424.

14

Politechnology

Manners Maketh Machine

P.A. HANCOCK

Contents

14.1 Introduction

> Manners must adorn knowledge, and smooth its way through the world.
>
> **Lord Chesterfield**

There is an old English saying that "Manners maketh man." The original quotation is attributed to William of Wykeham* and is a simple aphorism that acknowledges the tacit assumption that for human beings to live and thrive collectively, there must be a degree of formalization about their social intercourse. Although the bedrock of such mutual behavior would seem to be the law, our actual sharing of common assumptions goes much deeper. For we must all learn the social conventions first through the mediation of perception, the beginnings of language, and subsequently complex conventions and naming

* "Manners maketh man." William of Wykeman (1324–1404). The aphorism is also the motto of Winchester College and New College, Oxford.

processes. These capabilities are assimilated at our parents' feet. They are the very stuff of maturation, and although the precise conventions are highly culturally bound, the learning process itself cuts across all human experience. Now, into our diverse human cultures come potentially intelligent machines. And in respect of their growth and maturation, we can ask two critical questions. First, to what degree do these nascent centers of intentionality and potential consciousness need to adhere to the principles of human social intercourse? Second, and equally as important, do these machine have a collective of their own which now bestrides all human cultures? If they do, to what degree do they act on us as agents of convergent human cultural evolution? In asking these questions we also have to face the logical antithesis to all human-centered technological approaches, viz, should machines necessarily be made to behave according to human-centered rules? Is this their best mode of organization? Although the latter query might be antithetical heresy to those imbued with the spirit of human factors, it is a proposition we should always be obliged to consider. Overall, this a rather daunting set of issues to face, and so let us decompose the challenges they present so that we can consider them in their various component parts.

14.2 Evidence of Machine Intentionality

> If we are polite in manner and friendly in tone, we can without immediate risk be really rude to many a man.
>
> **Arthur Schopenhauer**

All entities possess etiquette.* From a human perspective, all objects and items with which we interact prove more or less useful with respect to the goal we wish to achieve. Thus, we find them more or less "polite" and useful to our purpose according to our own lights. The vast majority of the entities that exist in our natural environments came into being with absolutely no reference to human beings at all. Thus, they are largely neutral in terms of usage. For example, the rocks around us can be used effectively as weapons and much

* It should be understood that here I approach etiquette in a somewhat different manner from the editors of the present text (see for example, Miller, 2008).

less effectively as modes of transport, but their human utility (and thus their etiquette from our perspective) is completely framed by our own perceptions and intentions. In more formal terms, the affordance between the human and the entity is unalloyed by any conscious shaping other than the intention of the immediate human user themself. If we are able to step away from our own anthropocentric perspective for just one moment, humans also are simply part of the environment, and they most frequently exhibit very poor environmental etiquette with respect to other entities, often destroying the world solely for their own gain. But we must surely absolve humans from singular blame because most living organisms seek to manipulate the environment for their own species-specific advantage also. However, as we can rarely escape our own egocentric assumptions for long, let us persist in our argument from a human perspective.

Etiquette is interactive. And so we must also ask about the various agents involved. With theological exceptions, humans see themselves as the quintessential centers of intentionality. While we recognize goal-directed behavior in other animals it is rare that we defer to their intentions in any circumstances. Our etiquette with other animals being almost ubiquitously a master–slave relationship.* All living organisms evolve but, in adhering largely to Darwinian evolution, most do so very slowly, at least from the temporal perspective of a human observer (Hancock, 2007a). Machine evolution, in contrast, is largely Lamarckian in nature. It progresses at a much faster rate, which is evidently observable to even the dullest of mall-circling minds. But are machines now our partners and will they ever be our equal, or potentially our master? When we consider etiquette (which seems to be rather an effete way of saying manners), we largely consider it to be about interaction among equals. Although I would stand up when a lady entered the room, being now somewhat old-fashioned in my manners, I do not think I would do as much for her dog.† Among other things, manners seem to be a fine-grained sorting mechanism directed to distill who is *primum inter pares* (first among equals). For

* I am aware of the appalling behavior of some pet owners in indulging their little "friends" and even the idiocy of the few who leave vast fortunes to their surrogate offspring, but I am looking to the general not the particular here.

† If, however, her dog were a very angry Doberman Pinscher, I would be sorely tempted to make an exception.

example, one rarely thinks of etiquette as being between oneself and one's own small infant but rather it is a system of behavior among a set of strangers who nevertheless are still assuredly members of one's own group or caste.* And from this viewpoint, etiquette seems very much more for the upper classes. In English Victorian terms, manners wasted on the lower classes who are doomed to the hovel and the gutter—and are thus seen as representing a differing order of being altogether. Consequently, unlike law, etiquette can appear elitist and even divisive (since those who don't know or don't adhere to the rules of etiquette are readily excluded from subsequent discourse).

How do machines fit into this picture? I have said earlier that machines can now appear as centers of intentionality. By this I mean that all machines carry the imprint or shadow of their designer's original intent and, as any one machine will almost certainly have many designers (i.e., for their central processing units, visual displays, software, etc.), these intentions can appear more complex and disembodied as though they were of the machine itself and not its human creators. But this is not true at present. We have no current unequivocal evidence of independent machine intelligence and so the etiquette we are considering at this stage is fundamentally a human-to-human communication mediated by an evermore complicated information channel. And it is clear to anyone who has held a maddening phone soliloquy with an automated answering system that this communication often breaks down. Not to press the readers' patience too far, but we must consider that if and when machines do achieve this state of independent formulation of intent, their foundation will be that which we are now laying down. If, in their first incarnation of sentience, they treat us with the arrogance and disdain exhibited by many current systems, we are in trouble indeed!

Largely the problem of current machine insensitivity seems to be due to numbers and the inertia in time and space. In its traditional incarnation, etiquette is for the favored few who have the time and resources to engage in often byzantine and archaic behavioral

* I am very much reminded of Paul Scott's *Jewel in the Crown*, and the manners of the British Raj, or Robert Van Gullik's Judge Dee and the strict caste system of early China.

formalizations. One might even say that etiquette is a characteristic of decadence. But the modern world does not permit such niceties. The world is ever more crowded, ever faster. Machines are now not the luxurious privilege of the leisured few but are the instrument of labor of the working masses. Those of a certain "cake-eating" persuasion might argue that etiquette is wasted on the rabble! So let us talk more of manners and of law where here, the convention of law represents the lowest common denominator. And here we do have "laws" for machines. For example, Isaac Asimov (albeit within a science fiction realm) formulated the three basic laws of robotics. These are:

1. A robot may not injure a human being or, through inaction, allow a human being to come to harm.
2. A robot must obey orders given to it by human beings, except where such orders would conflict with the first law.
3. A robot must protect its own existence as long as such protection does not conflict with the first or second law.*

Unfortunately, it was after only a very few years of their development that robots broke all of these rules and indeed they continue to do so today (Hamilton and Hancock, 1986). At least we might expect that a machine would have the good manners not to destroy its user but this is also evidently not so. To cut a long story short, machines are mostly mannerless, human-law breakers and thus the equivalent talk of etiquette today might be analogized to consideration of the niceties of afternoon high-tea in the most secure wing of San Quentin, Sing-Sing, or Dannemora! It seems that our conception of interaction is then very much predicated upon our social assessment of the user. Thus, the machine is a lens through which we view ourselves. Disparate in time, space, and station in life, the way our machine

* I am assured by Wikipedia that these laws first appeared in Asimov's 1942 short story "Runaround" (caveat emptor). By the way, it is extremely evident that these "laws" are entirely human-centric and are all about protecting and serving human beings with no consideration of the desires and wishes of the poor robot. If you here replace the word "robot" with the word "slave" and then replace "human being" with "master" you can see how objectionable this might be. It further reminds us of the trial of Mr. Data on *Star Trek: The Next Generation* and his appeal to be considered of equal value. It is again noticeable here that the anthropomorphized Mr. Data wins his appeal while the Enterprise's main computer has to docilely and unquestioningly continue to obey.

reacts to us is an index of our own perceived value by the greater community. In case you think I am mistaken, be assured that the very rich and powerful never wait. Particularly, they never wait for a machine. Whenever he phones someone, you don't find that the President is put on hold and you don't see a four-star Army general navigating a complicated and poorly designed menu sequence of a commercial voice recognition system.

14.3 A Machine Culture

> Manners are of more importance than laws. Manners are what vex or soothe, corrupt or purify, exalt or debase, barbarize or refine us, by a constant, steady, uniform, insensible operation, like that of the air we breathe in.
>
> **Edmund Burke**

It may be some small crumb of comfort for the reader to know that the frustrations and annoyances that are attendant upon current computer interaction (for after all, the computer is what we are talking about) within the Western world are just as prevalent across the other cultures of the world. There are equally as many unhappy humans yelling at insensitive, uncaring software in Teharan, Buenos Aires, and Kyoto as there are in Washington, London, or Paris (and see Hoffman, Marx, and Hancock, 2008). And all of these people themselves form a culture. That is, they share these common behavioral reactions to these poorly designed and ineffective systems.* And each of these interactive failures represent evidence of bad manners but with one vital difference. At present, there is virtually no instant feedback mechanism to mark the incident and correct the behavior. True, there are large scale metrics of dropped calls or unsatisfied customers† but these do not provide the subtle, nuanced, and immediate response which can amend the immediately following behavioral act. In fact,

* With respect to the annoying and, indeed, horrible experience of voice-recognition menu systems, with which I have had yet another unsatisfactory encounter this morning, there are even Web sites that can tell you how to secure the most effective button-pressing sequence to get through to a live human being!

† Parenthetically, the automated system to register complaints is even more invidious than the original system being complained of. That's often because the complaint system is pawned off on the poorest software designer!

most current systems never learn or evolve themselves; the learning (if any) at present resides in the human designers. Thus, the next stage of machine evolution is for what I have termed "evolent systems," and manners may well be encoded in such machines as nested and hierarchic feedback loops.

Is the current machine culture then one of human misery? I would like to say so, but sadly for us aging "shock-jocks" of science, this is not so. By and large the work gets done and although users show signs of stress and distress, the whole edifice moves forward with some facility. And we can state, with some temerity, that this is a common machine culture across almost all of human society. It is therefore a source of sadness that the vestigial aspects of some of those global cultures use the burgeoning machine technology for something less than constructive aims (and see Hancock, 2003a).

However, can a machine culture be a unifying force? That is, can the communication possibilities that our common use of technology opens up be used as a positive source of integration? Assuredly so. But in this circumstance the machine might have to become the arbiter of etiquette. As manners vary across the globe, the machine as a partner may be able to provide a smooth conduit for intent by explicit recognition of the human etiquette rules of both sender and receiver. In this manner, the machine can become more than a language translator and can carry the full spectrum of behavior while filtering it through the traditional etiquette of each culture. Perhaps C-3PO will have a future after all. If technology is to act as a unifying force and an independent culture, it argues that we should not always adapt the machine exactly to the profile of the immediate user. For it argues that the machine has functional responsibilities to more than the person whose immediate hands are on the keys. It brings me to my final point about assumptions associated with human factors and human-centered computing.

14.4 What Machine Doth Manners Make?

> Good manners are made up of petty sacrifices.
>
> **Ralph Waldo Emerson**

If we continue to see the computer as an insensate tool that is solely designed for the gratification of human purposes, then tenets of etiquette mean that the technological systems should always be a dumb servant. At worst an abject slave, at best a Jeevesian form of avuncular butler, the machine is only ever an extension of individual and collective human will. This is not formalized behavior between equal strangers; this is the persistence of indentured servitude. And largely this must be so at the present time because almost any indications of intentionality on behalf of the machine are not self-generated but merely the ghosts and shadows of designers' intent, often misplaced far beyond the original context of conceived operations. What we are missing from this equation is evidence of original machine intent and machine learning. In Neal Stephenson's *The Diamond Age*, the intelligent book acts to educate the young human female ingénue, Nell. It is her guide, her mentor, and the source of her growing wisdom. It adapts to her state of development. It acts as an almost perfect adaptive human–machine system. But largely we do not get to see what the book itself wants. How does it grow? What satisfaction does it get from educating its human companion*

Humans and computers are not strangers. Rather, they are intimately familiar family members. The relationship is not a formalized one, as when meeting someone of the same rank from another part of one's own country. Rather, it is the relationship between a parent and a child. Right now, our computer children are sadly limited. They react, they respond, and there are tantalizing glimpses of an evident intelligence of sorts. But we cannot communicate as we would like to, and our progeny is not showing that nascent self-determination that will eventually characterize a full, independent adult. And that is certainly our fault, if fault is how one would wish to characterize it (Hancock, 2007b). What we need to achieve to bootstrap this form of self-determination is an intrinsic and express focus on machine self-learning. I have called such systems "evolent" in nature, but they demand the necessity to rethink some of the basic principles and assumptions of computer architecture (Hancock, 2007b) and a much greater focus on the role of emotion in the development of self-determination (and see Picard et al., 2001). However, we must always be very wary. Greek myths are redolent with stories of children who

* Notice how the context of this last sentence changes if we swap the respective words, owner or pet for the word companion.

murder their parents, and the theme of the new replacing the old is one that is common across all human cultures and all living organisms. We shall indeed reap what we sow, so we must sow mindfully.

14.5 A Machine Primer for the Care and Use of Human Beings

> It is possible for a man wholly to disappear and be merged in his manners.
>
> **Henry David Thoreau**

Is there a brighter vision of how humans and machines might get along together in the future? Of course there is. One only has to go to the latest "feel-good" sci-fi Hollywood extravaganza to see insipid and subservient C-3POs chastise childish yet powerful and efficient siblings such as R2-D2 on their necessary servility to their flawed but heroic human masters. These characters each embody the myth of the "happy slave," which has been swirling about Western civilization since the times of Aristotle, Plato, and Socrates. However, if we are to truly seek to inculcate intelligence into machines, then we will have to provide them with both motivation and with the opportunity to evolve. That they will do so not with a bang nor a whimper but in the blink of a human eye would then be almost inevitable and, in so doing, we shall have sown the seeds of our oppression or extinction. The fact that we now seek to create such servants and to chase the vision that their nascent effect will be uninfluenced by the constraints we will seek to place upon them merely perpetuates some of our own human illusions. And here I was trying to be optimistic!

Perhaps, indeed, when the balance of power tips, our mechanical children will, as Moravec (1988) opines, be too busy with the greater spaces of the universe to bother too much with their human forebears. Perhaps they will spare a few of their trillions of machine cycles per second to create virtual paradises for their carbon progenitors after the manner of Captain Christopher Pike in the first ever Star Trek episode. Perhaps we shall eventually live in their illusions. Alternatively, we could end up as the most entertaining exhibit in their zoo. I would like to be able to envisage a human–machine social utopia, but it is almost beyond me to do so. Almost, because St. (Sir) Thomas More's original Utopia was nowhere near the Nirvana it has always been upheld to be by those who have not read his original work (More, 1516).

14.6 Summary and Conclusion

> The test of good manners is to be patient with bad ones.
>
> **Solomon Ibn Gabirol**

As Homer Simpson is being pounded by Drederick Tatum in his putative challenge for the 'Simpsons-Cartoon-World" boxing title, the background music plays the lament "why can't we be friends?"* At its heart, etiquette is about some degree of give and take. However, human-centered approaches are not about give and take, they are about take and take. We currently operate on the tacit, and often frequently explicit, assumption of the ascendency of the user. I think this assumption has to be constantly questioned. First, unlike the apparent assumptions of usability, each individual does not always know exactly what is best. Thus, customizing systems and making them react according to the wishes of each individual may serve the tenets of individuation (Hancock, 2003b; Hancock, Hancock, and Warm, 2009); however, it does not always ensure optimal operations. When the use itself is idiosyncratic and personal (as many PC uses are, almost by definition), the suboptimization is a matter of personal account. However, when the use is social and with responsibilities to the greater whole, then merely pleasing the individual with his or her hands presently on the keys is insufficient. Here, where the user may make mistakes that will cost others their lives, the computer has the right to be ill-mannered and even downright rude. Asimov would say it has the moral imperative to do so. Thus, if we are to further the metaphor of etiquette as our interactive template with machines, we will have to concede something of ourselves and, at a latter point in time, when machines begin the realization of sentience, we may well have to give up much more. Are we ready to make these sacrifices? From personal observations of the egocentricity of the human species, I would say probably not. And yet the inexorable progress of

* In a highly appropriate way, the song is sung by Smashmouth.

technology, on which we become evermore dependent, may give us no choice. We may well be the architects of our own demise. But if etiquette is to the fore, at least the machines may say please before they ease us along the well-worn path to extinction.

Acknowledgments

The present work was facilitated by a Grant from the Army Research Laboratories under ARL, Task Order 81: Contract DAAD19-01-C-0065; "Scalable Interfaces for Dismounted Soldier Applications," Mike Barnes, Grant Monitor. The views expressed here are those of the author and do not necessarily represent those of any named individual or agency. I'd very much like to thank Chris Miller and Caroline Hayes for their kind invitation to contribute to the present text. I should emphasize that their comments on the present chapter and their prior works have been most helpful in creating and revising this work. Although we may disagree as to aspects of the present issue, they are true scientists in eliciting and including works that do not necessarily accord directly with their own views. I am grateful for their tolerance and insight.

References

Hamilton, J.E. and Hancock, P.A. (1986). Robotics safety: Exclusion guarding for industrial operations. *Journal of Occupational Accidents, 8*, 69–78.

Hancock, P.A. (2003a). The ergonomics of torture: The moral dimension of evolving human-machine technology. *Proceedings of the Human Factors and Ergonomics Society, 47*, 1009–1011.

Hancock, P.A. (2003b). Individuation: Not merely human-centered but person-specific design. *Proceedings of the Human Factors and Ergonomics Society, 47*, 1085–1086.

Hancock, P.A. (2007a). On time and the origin of the theory of evolution. *Kronoscope, 6* (2), 192–203.

Hancock, P.A. (2007b). On the nature of time in conceptual and computational nervous systems. *Kronoscope, 7*, 1–12.

Hancock, P. A., Hancock, G. M., and Warm, J.S. (2009). Individuation: The N=1 revolution. *Theoretical Issues in Ergonomic Scince, 10* (5), 481-488.

Hoffman, R.R., Marx, M., and Hancock, P.A. (2008). Metrics, metrics, metrics: Negative hedonicity. *IEEE Intelligent Systems, 23* (2), 69–73.

Miller, C.A. (2008). Social relationships and etiquette with technical systems. To appear in B. Whitworth and A. de Moor (Eds.), *Handbook of Research on Socio-Technical Design and Social Networking Systems*. Information Science Reference; Hershey, PA. In submission.

Moravec, H. (1988). *Mind Children: The Future of Robot and Human Intelligence*. Harvard University Press: Cambridge, MA.

More, T. (1516). *Utopia*. (and see http://oregonstate.edu/instruct/phl302/texts/more/utopia-contents.html).

Picard, R.W., Vyzas, E., and Healy, J. (2001). Toward machine emotional intelligence: Analysis of affective physiological state. *IEEE Transactions on Pattern Analysis and Machine Intelligence*, *23* (10), 1175–1191.

15

EPILOGUE

CAROLINE C. HAYES AND
CHRISTOPHER A. MILLER

Etiquette is a lens through which to view interactions between humans and computers, or more generally, between humans and machines. Etiquette is essential in all cultures to facilitate the establishment of trust, develop relationships, and support effective and productive interactions with other people. Etiquette is equally essential for enabling people and computers to do the same. When we interact with other people, it deeply engages cultural, psychological, neurological, and physiological aspects of our make-up; perhaps it should not be surprising if similar phenomena are evoked when we interact with computers. The chapters in this book have explored etiquette in many applications ranging from computer tutors that train people in the language and etiquette of other cultures, to computer coaches that encourage people to exercise more, to robots that assist patients in the hospital. They explore many facets of human–computer etiquette, including:

- The potential for cultural clashes when people from many cultures interact or when people from one culture interact with a computer interface designed by another
- The impact of politeness on multiple aspects of human performance: patient compliance, student learning outcomes, navigation efficiency, etc.
- How to create computer agents and robots that are capable of politeness through speech, gestures, gaze, interruption strategies, and other methods
- The potential for a computer's etiquette to influence the emotional responses, trust, and long-term relationships people form with them

- New approaches that may allow computer agents and robots to better understand people by monitoring their brain waves, heart rate, facial expressions, tone of voice, and other indicators
- The potential for polite or rude software design to influence behavior of society as a whole, and our future

Through these chapters it has been our goal to illustrate how human–computer etiquette is an important aspect of human–computer interaction (HCI), and to show that it can strongly impact the effectiveness of those interactions. However, up until now, this issue has received little attention and has gone largely unrecognized. This is because software developers, HCI researchers, and people in general do not tend to think of computers as objects to which etiquette rules should apply. Furthermore, because etiquette conventions are often implicit, it may never occur to people when they become irritated with a computer that their irritation might be the result of a mismatch between their etiquette expectations and the computer's etiquette behaviors (or lack thereof) as programmed into the software by the designer.

The second goal of the book has been to make software developers aware that (1) users *do* indeed have etiquette expectations of computer agents, and those expectation may be powerful and implicit; (2) those expectations are highly nuanced and may vary greatly, depending on the roles that humans and computers play in an interaction, the specific task context, the culture of the user's work discipline, organization, and nationality; and (3) that human–computer etiquette can have a large impact on the frustration levels, performance, and safety of the people that use the software, and its ultimate success or failure. We recommend that software developers should be given explicit training and education that will enable them to recognize the implicit etiquette expectations embedded in their own culture. These must be made explicit in software, and the differences spelled out clearly in the etiquette expectations between different cultures. The latter is especially important if the software is aimed at a diverse world market, or at a specific market embedded in a culture different from the developer's own. At the very least, designers should include an "etiquette perspective" as an important part of the design process in which they consider how the computer will be perceived as a social actor by its users.

The third goal of the book has been to offer human–computer etiquette as a perspective that can explain many of the emotional reactions that people have to computers—reactions which affect how they perceive and use computers, and what benefits they derive from them. This perspective can provide researchers with frameworks and models with which to explore, understand, measure, evaluate, and explain how and why people react to their computers.

For software developers, the "etiquette perspective" provides them with another approach to improve the user's experience with, and emotional reaction to, their products. In the current economic climate of intense global competition, product developers are acutely aware that to be competitive, products must not only function effectively on a technical level, but must also be easy and enjoyable to use. Additionally, it provides designers with a way to understand that software tools designed for customers in emerging markets, such as India and China, may need to interact with customers differently from similar tools designed to satisfy customers in markets such as Europe. An explicit understanding of human–computer etiquette can provide designers with an approach through which they can better support peoples' differing ways of working and living effectively and pleasantly with their computers.

This book provides an introduction to the concept of human–computer etiquette, but much remains to be explored and understood. We have only begun to answer the questions introduced in Chapter 1; they form the basis of a research agenda for future investigations in human–computer etiquette. Computers have often been viewed as lacking in nuance and sensitivity to context; the theory and practice of human–computer etiquette may be exactly what is needed to advance computers into the next age of computation: one in which computers are intuitive, responsive, and appropriate, where people form long-term relationships with them, and where they are team players and possibly even members of society. As we increase our understanding of human–computer etiquette we will likely learn as much or more about ourselves as social beings. Finally, computers are here to stay so we had best learn to get along with them.

Index

B

C

U